DURBAN'S CLIMATE GAMBLE

DURBAN'S CLIMATE GAMBLE

Trading Carbon, Betting the Earth

Edited by Patrick Bond

Contributors

Ashwin Desai

Joel Kovel

Larry Lohmann

Oscar Reyes

Khadija Sharife

Yash Tandon

Goolam Vahed

Del Weston

UNISA PRESS

PRETORIA

First edition, first impression

ISBN 978-1-86888-685-2

Published by Unisa Press
University of South Africa
P O Box 392, 0003 UNISA
Printed byHarrys Printers

Project manager: Ingrid Stegmann
Book Design & Typesetting: Designwave
Editor: Ellen Cameron
Proofreading: Ellen Cameron, Marjorie van Wyk & Natalie Bosman
Indexer: Mirie van Rooyen

Contents

Acknowledgements

This book began as a collaboration with the US-based journal *Capitalism Nature Socialism (CNS),* and is an expansion of its December 2011 issue, which is entirely devoted to COP 17 and Durban. In addition to Ingrid Stegmann, who did a great job as project manager for Unisa Press, we wish to express particular thanks to Karen Charman, Managing Editor of *CNS* and to *CNS* reviewers and editors.

About the Contributors

Patrick Bond is Director of the Centre for Civil Society and Professor in the School of Development Studies at the University of KwaZulu-Natal. His books include *Politics of Climate Justice* (2011); *Climate Change, Carbon Trading and Civil Society* (2009); and *Unsustainable South Africa* (2002).

Ashwin Desai is a professor of sociology at the University of Johannesburg. He is the author of numerous books including *We Are the Poors: Community Struggles in Post-Apartheid South Africa* (2002); *Blacks in Whites: A Century of Cricket Struggles in KwaZulu-Natal* (also 2002, co-authored with Vishnu Padayachee, Krish Reddy & Goolam Vahed); *Inside Indenture: A South African Story, 1860-1914 (2007); The Race to Transform: Sports in Post-Apartheid South Africa (2010).*

Joel Kovel is editor emeritus of the journal, *Capitalism Nature Socialism (CNS)* and the author of numerous books including *Overcoming Zionism: Creating a Single Democratic State in Israel/Palestine* (2007), *The Enemy of Nature: The End of Capitalism or the End of the World?* (2002), *Red Hunting in the Promised Land: Anticommunism and the Making of America* (1997), *History and Spirit: An Inquiry into the Philosophy of Liberation* (1991), *The Radical Spirit: Essays on Psychoanalysis and Society* (1990), and *White Racism: A Psychohistory* (1970).

Larry Lohmann is a researcher and activist based at a British NGO, The Cornerhouse. He is co-author, with the late Ricardo Carrere, of *Pulping the South: Industrial Tree Plantations in the Global Paper Economy* (1996) and has published articles in *New Political Economy*, the *Bulletin of Concerned Asian Scholars*, *Science as Culture*, and many other journals.

Oscar Reyes is a researcher with Carbon Trade Watch in Barcelona, environment editor of *Red Pepper* magazine, and co-author of *Carbon Trading: How it Works and Why it Fails.*

Khadija Sharife is a journalist, researcher, and visiting scholar at the Centre for Civil Society based at the University of KwaZulu-Natal. She is the regional correspondent for *The Africa Report* magazine and author of *Tax Us If You Can Africa*.

South Durban Community Environmental Alliance is a coalition of community and environmental organizations launched in 1995. Its members struggle together for environmental justice in heavily polluted south Durban and beyond. They seek a more sustainable Durban where the health and wellbeing of people and the natural environment are ensured.

Yash Tandon recently served as Executive Director of the South Centre, an intergovernmental think tank of the developing countries. His long career in national and international development spans time as a government minister and policymaker, a political activist, a professor, and a public intellectual. A native of Uganda, he was deeply involved in the struggle against the dictatorship of Idi Amin and received his degrees in economics and international relations from the London School of Economics. He has taught at Makere, and Columbia University. He was Founding Director of the Southern and Eastern African Trade Information and Negotiations Institute. He has authored more than 100 books and scholarly articles.

Goolam Vahed is an Associate Professor in History at the University of KwaZulu-Natal. He has published widely on the colonial encounter between Indians, Africans, and whites in South Africa; the transformations of Indian identities; Islam in South Africa; and the role of sport and culture in South African society.

Del Weston is a doctoral student at Curtin University, Australia.

Foreword

A Gathering Storm

By Joel Kovel

Sixteen times the ponderous UN climate bureaucracy has cranked out an annual 'Conference of the Parties' in the vain hope of finessing the menace of climate change while making a lot of money at the same time. The last two years have signalled the beginnings of the inevitable breakdown of the doomed process: in 2009, COP 15 in Copenhagen proved a prodigious betrayal – or was it the logical implication? – of the Kyoto protocols laid down in 1997, while 2010's COP 16 in Cancún will be remembered as the beginnings of a breakthrough by long-suppressed voices of the indigenous and peasantries – a rising of the South against the regime of carbon-based capitalism that is literally destroying the human world.

Now, in 2011, it appears that not just the climate is hotting up, but the level of struggle as well. As I write, Via Campesina has pledged to appear in force, bringing subsistence agricultural producers from all over the continent of Africa, and perhaps from elsewhere as well, to join in protesting the disaster that is the UN COP regime. There is no telling how many will join them from urban settings across the continent and around the world. But there is much to be told about where they will be joined – the city of Durban, capital of the province of KwaZulu-Natal and host to COP 17. It is the purpose of this book to tell the story of resistance to COP 17 through the lens of its locale.

Like most ex-colonial settings, Durban bears a full set of scars inflicted by capital's war on humanity and nature. But it also glows with the flame of struggle, from Gandhi's development of *Satyagraha,* to the apartheid era, and into its neoliberal sequel. This latter phase has been especially vicious given the heavy price exacted by capital for liberation from the chains of white racism. The post-apartheid years, in South Africa generally and Durban particularly, have been a race to the bottom measured in terms of aggressive penetration of ecosystems for the sake of accumulation. All of these are connected to the carbon complex in one way or another, and all of them enter into, and violate, the life-worlds of the poor. And all are ardently contested in the spirit of Durban, from popular, neighbourhood-based movements, to the University of KwaZulu-Natal's Centre for Civil Society under the direction of Patrick Bond, who has been the chief architect of this work.

And so the 'Conference of Polluters' is to be confronted by the multitude of those who have suffered pollution. This creates the need for our guide, which will situate COP 17, to lay out the story of the pernicious carbon regime, detail the political geography of Durban, draw forth the lessons of these tormented times, and present the argument for inspired resistance. The point of view argued in this volume makes no bones about which side it is on, and disdains a hypocritically 'balanced' view. The interested reader may find copious amounts of official text to present what the establishment wants you to think. Our guide is written, rather, about, of, by, and for, those on the underside, which is to say, the vast majority of earth's inhabitants.

The confrontation radiates outward in space and time. All Africa will be watching these proceedings and hopefully sharing in resistance – as may the South at large, and the metropolitan regions. As for the propitiousness of the time, it is hard to escape the impression that the year 2011 represents a kind of sea change in the emergence of radical struggle across the globe, whether as Arab Spring, outbreaks in the EU against draconian cutbacks, urban uprisings in Britain, even a million students in Santiago, Chile, thirty-eight years after the murder of Allende, singing *Un pueblo unido jamas sera vencido …*

There have been stirrings, too, in the most retrograde areas, such as Israel and the United States (where, as of this writing, hundreds have been arrested at the White House protesting the transfer of tar-sand petroleum from Alberta to the Gulf of Mexico). Each occasion is uniquely shaped by local factors. But all, and those to come, are ultimately responses to the great crisis of capital, writhing between its accumulation and ecological aspects. Ruling classes tend to become more vicious as their bell tolls. They crack down harder, exploit more nastily, intrude and surveil more recklessly, and throw to the winds the softening features of liberal-democratic society, shedding legitimacy as they go and releasing increasing numbers of people, whether the young who face dead-end lives or the desperate peasantries, to undertake more extreme measures. We see the forthcoming COP 17 events in such a context, and offer this collection as a contribution to their struggle, the greatest challenge ever faced by humankind.

Introduction

Welcome to Durban

Patrick Bond

This book is devoted to exploring two interlocking, overlapping scales of political ecology: local eco-social conditions and environmental justice campaigning in Durban, South Africa and climate justice advocacy against market-based 'false solutions' at the global scale. Their fusion at the 17th Conference of the Parties (COP 17) to the UN Framework Convention on Climate Change (UNFCCC) links up myriad battles over ecosystems that mainly revolve around 'neoliberalised nature'.

For example, the pricing of socio-ecological services has been extreme when applied to carbon sequestration, and Durban is a leading site. More generally, the tendency to commodify nature has become the defining philosophical stance behind global environmental governance, with inevitable conflicts, of which Durban's are emblematic. For example, 'cost-reflective' (unsubsidised) pricing of South African electricity, at a time of increased coal-fired power plant construction, has generated an intense battle over access to energy in Durban's informal settlements, where illegal connections are rife. The struggle over the price of water and sanitation has led to South Africa's famous water wars. The 'green economy' (often interpreted as multinational corporate promotion of biofuels and geoengineering) and 'Payments for Environmental Services' are set to become conventional wisdom at the Rio+20 summit in mid-2012, but not to ensure that the North's ecological debt to the South is properly acknowledged, but rather to establish a range of financialised investment options that securitise 'natural capital'.

In contrast, our contributors grapple with how to advance 'climate justice', the name of the new bottom-up global movement tackling the greatest challenge in human history.

After all, from above, *failure* is the only way to summarise 16 years of talk by United Nations negotiators from national states (influenced by fossil-fuel-dependent capital, neoliberal multilateral agencies, and the big environmental NGOs from Washington and London). That failure is programmed to continue in Durban, with only climate finance and technology on the agenda, as sites where a deal may be made. Yet so far, money to make it happen is scarce.

As the world will learn, South Africa also exemplifies climate *in*justice. Vast amounts of CO_2 emissions come from giving the world's cheapest electricity to the world's biggest mining and metals corporations (such as BHP Billiton and Anglo American), which in turn permanently degrade water resources and pollute the air. In April 2010, the World Bank made a US$3.75 billion loan to Eskom (for the Medupi coal-fired power plant) to feed the greed of coal-burning capital, along with payola for the ruling party, while poor people suffer repeated annual 25 percent electricity price hikes to repay the loan. Vast opportunities for climate jobs and renewable energy in SA have been nearly entirely ignored. Meanwhile, electricity disconnections affect millions of low-income people each year, leading to intense 'service delivery protests'. And South African President Jacob Zuma was one of the five leaders who signed the unfair, unambitious and non-binding Copenhagen Accord in 2009.

Yet, from the standpoint of climate justice, the choice of the International Convention Centre (ICC) in Durban, South Africa to host COP 17 from November 28 to December 9, is intriguing: some of the country's most intense environmental, social, and labour struggles are underway here.

Climate justice for Durban?

Welcome to Durban, a city of 3.5 million people (with a racial division of 63 percent African, 22 percent Indian, 11 percent white, and 3 percent coloured) living within 2 297 square kilometres. Durban's formal metropolitan identification, eThekwini, is regularly used in official discourse, but hardly ever in public. Sadly, the colonial legacy continues: the governor of the Cape Colony, at the time settlers planned the city in 1835, was Lieutenant General Sir Benjamin d'Urban. The replacement name 'eThekwini' was publicly rejected even by the three-term post-apartheid mayor, Obed Mlaba, as in the isiZulu language it means 'a bull's testicles', a phrase associated with the geographic appearance of Durban's Bluff and port from a prominent ridge, where King Shaka trained his army (de Boer 2007). In spite of many controversial street-name changes in the late 2000s, the local rulers did not sort out the overall designation, so unlike many official foibles, the name eThekwini remains an uncorrected post-apartheid *faux pas*.

Welcome to Durban, where activist critiques of local and global elites are well-rehearsed. The World Conference against Racism was held here in 2001, and at least 15 000 people demonstrated against the UN at the ICC, because thanks to Bush regime pressure, UN Human Rights Commissioner, Mary Robinson and host Thabo Mbeki, refused to put onto the agenda either Israeli racism against Palestine or Western reparations for slavery, colonialism, and apartheid. (On the latter, Mbeki rebuffed even African heads of state, including Nigeria's Olusegun Obasanjo [Bond 2006]). Exactly a year later, at the Johannesburg World Summit on Sustainable Development, 30 000 activists protested UN commodification of nature: 'bluewashing' of corporate polluters through the Global Compact, 'Type II Partnerships' with multinational capital, neoliberal social and economic policies, and UN-endorsed water privatisation and carbon trading (Bond 2006). Other protests were held regularly at the Durban ICC against World Economic Forum sessions and Pretoria's neoliberal 'New Partnership for Africa's Development' (when the African Union replaced the Organization of African Unity in 2002).

There is a rich history of international eco-social solidarity for climate activists to draw upon. South Africans and global allies fought racial apartheid and won in 1994; activists demanded access to anti-retroviral HIV/AIDS medicines in the late 1990s, defeating Big Pharma, the World Trade Organization and the Bill Clinton and Thabo Mbeki regimes by 2004. Today, more than a million South Africans have access to medicines that, a decade earlier, cost US$15 000 per person per year. Indeed, the first high-profile martyr of this cause was Durban's Treatment Action Campaign educator Gugu Dlamini in Umlazi in 1998, and the first sustained critiques of Mbeki's AIDS denialism occurred in 2000 at the XIII International AIDS Conference in Durban.

Political history was made here with the initial resistance by Zulu King Shaka to British settlers in the early 1800s, and a century later with Mahatma Gandhi's (1928) political innovation, *Satyagraha* (truth-force civil disobedience, used to advance Indian ethnic rights). South Africa's modern trade union movement began here in Africa's largest harbour with the 1973 dockworker strike. The rise of the country's 'new urban social movements' began in the Chatsworth community in 1998 (Desai 2002). Over the past fifteen years, South Durban activists have led the way on environmental justice activism, with the closure of the Umlazi landfill – a major 1996 victory for activists headed by Bobby Peek (who subsequently founded the Friends of the Earth affiliate, groundWork). Shack dwellers emerged as a major movement, when in 2005, Abahlali baseMjondolo was founded at Kennedy Road (Kell and Nizzo 2011). Nearby, at the Bisasar Road landfill, a five-year long campaign was waged by Sajida Khan (until her death in 2007) against South Africa's first big Clean Development Mechanism (CDM) (Bond, Dada, and Erion 2009), which in turn led to the founding of the Durban Group for Climate Justice international network in 2004.

Activists in other major South African cities also host ongoing protests on climate, energy and related issues. For example, Earthlife Africa and the Anti-Privatisation Forum regularly engage in actions at the headquarters of the vast state electricity company, Eskom, and the coal-to-oil multinational Sasol – South Africa's two largest CO_2 emitters. And in mid-2011, Greenpeace dumped loads of coal outside Eskom's front door to protest its fossil fuel addiction. Cape Town, and the Eastern Cape surfing village, Jeffreys Bay, have periodic anti-nuclear demonstrations. South African activists founded Climate Justice Now! SA in 2009, and joined with mainstream groups (even the pro-corporate World Wildlife Fund) in a 'C17' network of civil society activists. Their aim: to successfully host an alternative summit (People's Space) at the Durban University of Technology (DUT), including a mass march on December 3, 2011 from the Curries Fountain Sports Ground (next to DUT) through Warwick Junction's market, up to the US Consulate and City Hall, past the ICC, and ending several hours later at the Indian Ocean for a 'Going Away Party for the Beach' concert and protest.

Among the Durban-based groups promoting a stronger political message, within the broad tent, are SDCEA, groundWork, the Centre for Civil Society, the ECOPEACE Party and Democratic Left Front, Earthlife eThekwini, Timberwatch, the Diakonia Centre, Streetnet, Umphilo waManzi, and allied community, environmental and labour organisations. Their aim is to ensure that delegates to the COP 17 get the message: cut a genuine deal or face South Africa's famed social protests, which are among the world's most prolific. To make the point, various sites across Durban will illustrate climate contradictions and social conflicts.

Sites of significance for climate activists abound, beginning in the north at the King Shaka International Airport. For example, the source of Durban's main water supply is the Inanda Dam, whose methane from rotting vegetation is, like many tropical and subtropical dams, a substantial greenhouse gas source. Acquisition of land for the late 1980s dam is still contested by the Qadi people, who are victims of both displacement and a corrupt apartheid-era state relationship with a co-opted chief. Near the dam is the Phoenix settlement, where a century ago, Gandhi established a printing press and intentional community to further *Satyagraha* (which he took with him when he left South Africa in 1914, and which was responsible for ridding India of the British empire).

Due east across the main highway from the Phoenix settlement is the town of Umhlanga, where the continent's largest shopping mall, Gateway, exhibits luxury goods for those mainly white shoppers who, at the time of South Africa's democratisation in 1994, escaped the hustle and bustle ('crime and grime' in their words) of a rapidly integrating central Durban. Gateway, built on a rubbish dump that was closed in the mid-1990s, is the sixth largest electricity consumer in Durban. Further west is the notorious Cato Ridge site where Thor Chemicals dumped illegally imported mercury waste (whose residues now make Inanda Dam fish too toxic for human consumption) and the Assmang manganese smelter, which consumes a vast share of Durban's electricity and which has killed numerous workers in explosions and manganese poisoning.

Moving back east to central Durban, we soon encounter the harsh smells and perfume rods of the Bisasar Road dump and Africa's largest landfill, the subject of community protest when the World Bank prevented its closure in the early 2000s, in order to commodify the methane that the dump produces to sell to the emissions markets, as Khadija Sharife describes in her chapter, 'Colonising Africa's Atmospheric Commons'. Placed in the black Clair Estate neighbourhood during apartheid, it should have been closed when Nelson Mandela came to power, as African National Congress (ANC) pamphlets in the 1994 election promised the community it would be, but because of the methane-electricity investments, there is now no prospect for closure before it fills up around 2020.

Next to Bisasar is a shack settlement, Kennedy Road, once well-organised by the group Abahlali baseMjondolo before it was attacked in September 2009, resulting in the group being evicted and replaced by local ANC powerbrokers. But subsequent violent battles with municipal contractors responsible for disconnecting electricity in mid-2011 showed that the ruling party never quite gained hegemony in Kennedy Road. Abahlali, meanwhile, won a major Constitutional Court victory against provincial housing legislation that would have led to more forced removals. In early 2010, Abahlali held a major march for shack dweller rights in central Durban, despite efforts to divert the march away from City Hall.

From Bisasar, travelling east a few kilometres past the 'big box' retail stores to the end of Umgeni Road, travellers arrive at Warwick Junction, described by Bond and Desai in the first chapter. Next door is the DUT, host of the People's Space alternative summit (university chancellor Ela Gandhi is Mahatma's granddaughter and directs the Satyagraha Centre there). Continuing towards the sea, the arch of the Moses Mabhida Stadium is unmistakable, and a few blocks due south is the ICC. Two blocks further southwest are Durban's City Hall, the much-maligned central business district, and the US Consulate atop Old Mutual's main central Durban building. This gives US officials a commanding view of the Durban harbour a couple of blocks south; these consular staff were responsible for intensifying port security against counterfeit goods, and in the process, ridding the harbour of the 'security risk' of fisherfolk in 2004. Due west, atop the highest hill in central Durban, is the University of KwaZulu-Natal's (UKZN) Centre for Civil Society, occupying the top floors of the Memorial Tower Building.

From there we enjoy clear views east across the harbour to the Island View refinery, which is notorious for a fatal explosion in 2007, described in the SDCEA chapter 'Feeling the Heat in Durban,' in the pages below. Island View marks the beginning of Africa's largest petrochemical complex, intensely watchdogged by the SDCEA, whose traditional targets include the Engen refinery, Mondi paper, the Sapref refinery and the Toyota assembly plant. In 2009, a major climate summit was hosted by SDCEA, which also organised early protests against the World Bank's funding of the Medupi Power Station. Activism continues in South Durban – one of the world's most dangerous petro-chemical complexes – including one of the '1 000 Cancúns' protests in December 2010. Activists occasionally win, as the victories in the Chatsworth electricity struggles and the Bulbul Drive dump attest.

Durban's contradictions

The city is rife with contradictions and social conflict. Officials have failed to incorporate climate change in economic development planning, leaving citizens far more vulnerable than they should be. For example,

- Gambling that the city's future will be based on tourism, major sports events (the World Cup in 2010 will very likely generate an Olympic bid for 2024), and international trade is dubious, given what we now know about the need for national and global carbon taxes.
- Attempting to profit from climate change through CDM carbon trading gimmicks, at three landfill sites, is not only unworkable because of the repeated failure of the

emissions markets (and in the case of the La Mercy landfill, a technical failure), it is also immoral because it allows developed countries and their companies to maintain greenhouse gas emissions, while neglecting green alternatives for waste disposal at home.

- Failing to fund genuine climate jobs, public transport, and the just transition away from fossil fuel addiction, is another example of short-sightedness, among a string of stupid decisions and failed gambles that often involve public subsidies.

The list of bungled projects and missed opportunities includes the Point and ICC, the loss of Blue Flag status designating clean beaches, the attempted razing of the Early Morning Market, bus privatisation, mass electricity service disconnections, a ballooning housing crisis, evictions of fisherfolk from the port, non-consultative street renaming, and vast cost-overruns at the unneeded Moses Mabhida Stadium.

As a result of these contradictions, the Association of Science in South Africa (ASSAf) was commissioned to work with the municipality's climate adaptation office to produce a major report, *Towards a Low Carbon City: Focus on Durban* (ASSAf 2011), in August 2011. If they took the title seriously, new Durban Mayor, James Nxumalo, and an incoming manager (not yet chosen as of this writing in mid-2011) would be faced with a formidable challenge: Switching the city's economic strategies in the direction promised and facing up to the threat we face, especially the threat that Durban's major industries pose, as a result of runaway greenhouse gas emissions.

We needn't reiterate, here, concerns about future rising sea levels and the vulnerability of Durban's beachfront and valleys to extreme storms, as witnessed in 2007, 2008, and 2011 (flooding that will overwhelm dirty Durban's decrepit stormwater drainage system, landslides on our hilly terrain, droughts that draw new climate refugees from the region into what is regrettably a xenophobic populace, the disruption of food chains, and other coming disasters).

However, what might be termed South Africa's 'mitigation-denialism' deserves elaboration, since the ASSAf report is now a prime case. Not only did Planning Minister, Trevor Manuel announce in August 2011 that he expects the North to pay South Africa up to US$2 billion per year, through the Green Climate Fund he co-chairs – when in reality it is South Africa that owes a vast climate debt to Africa given our world-leading rate of CO_2 per GDP per person – but ASSAf seeks to persuade politicians that Durban can 'entrench its reputation as South Africa's leading city in terms of climate change actions [sic]'. This is pure hubris, because ASSAf's 262-page study shies away from any critical mention of *high-carbon,* Durban's unprecedented public subsidies for long-distance air transport, shipping, fossil-fuel infrastructure, highway extension and international tourism.

For example, the ASSAf study tells us nothing about the US$35 billion that 'back of port' planners have in mind for the South Durban basin. The project will displace residents of the 140-year-old Clairwood neighbourhood, to allow for further expansion of the harbour and its ships' dirty bunker fuel problem. It proposes a new highway leading to more container terminals and super-toxic petrochemical facilities, as well as a new pipeline (through black neighbourhoods) to Johannesburg that will double the amount of oil transported from the petrochemical complex. There is also a planned expansion of the automotive industry and construction of a huge new harbour on the old airport site. Yet, this project warrants no mention in ASSAf's report.

Further north, ASSAf says nothing about the damage done by building the US$1.2 billion King Shaka International Airport way too early, and way too far north of the city, nor (aside from a throwaway reference in the governance chapter) about the mostly empty US$450 million Moses Mabhida Stadium, built for the 2010 World Cup, next door to the city's world-class rugby stadium. All these mega-investments certainly make Durban 'South Africa's leading city in terms of climate change actions', but opposite the way ASSAf claims.

In a failure of analytical nerve, ASSAf scientists appear too intimidated to discuss these expensive mistakes in polite company, much less argue for a detox-rehab of Durban's carbon-addicted corporates. Yet it makes no sense to avoid the harsh reality of fast-rising emissions in sectors that make our city exceptionally vulnerable when carbon taxes do finally kick in, given how far Durban is located from the world's main markets, and given the adverse implications for tourism. At one point, buried in a dry table, are the names of Durban's biggest emitters, measured by consumption of *municipal* electricity: the Mondi paper mill, Sapref and Engen oil refineries, Toyota, Frame Textiles, and the Gateway and Pavillion shopping malls. But the city's biggest contributor to climate change, via the *national grid's* coal-fired power plants, is the deadly manganese smelter, completely forgotten in ASSAf's study, even though at the mining conglomerate, Assore's 2010 annual report concedes, 'Electricity consumption is the major contributor to Assmang's corporate carbon footprint and reflects energy sourced from Eskom grid supply, particularly by the Cato Ridge Works.' Nor in ASSAf's chapter on 'the national context' do we learn that South Africa is building the world's third and fourth largest coal-fired power plants, Eskom's Kusile and Medupi, with a US$3.75 billion loan from the World Bank, in spite of fierce opposition from civil society. Not mentioned, either, are apartheid-era Special Pricing Agreements that give BHP Billiton and Anglo American the world's cheapest electricity (US$0.02 per kilowatt hour), about one-eighth of what ordinary households pay. Nor is there a word about the millions of poor South Africans disconnected from electricity, unable to absorb the 130 percent price hike Eskom has imposed, since 2008, in order to service loans on the coal-fired generators.

These gaping holes are too wide for even Durban's most skilled greenwashers, like municipal climate adaptation manager, Debra Roberts, to hide. To her credit, she joked that 'You want to get me fired for publicly agreeing with you', and then did just that when at the ICC launch, I drew attention to these white-elephants-in-the-room. ASSAf chief executive, Roseanne Diab, replied that the city's main mitigation focus should be Durban's anarchic truck-freight transport mess, which she claimed can be tackled by air-quality regulation. That might be the case if South Africa had the USA's Clean Air Act (which considers greenhouse gases to be pollutants – something our South African Air Quality Act doesn't). And it might also help if the municipality had an effective air pollution monitoring unit, but in March 2011 it was stripped of most of its staff by Durban City Manager, Mike Sutcliffe.

Likewise, ASSAf enthusiastically promotes landfill methane gas-to-electricity conversion at Durban's Bisasar Road dump, without observing (as do most academic articles) that Africa's largest CDM is actually one of the world's primary cases of carbon-trading *environmental racism*, worthy of a front-page article in the *Washington Post* in 2005, on the day the Kyoto Protocol took effect (Vedantam 2005). In its enthusiasm for such financing, the ASSAf study also forgets that the COP 17 will witness the demise of Kyoto, the treaty that mandates these kinds of carbon-trade investments in places like Durban. Blithely, ASSAf scientists recommend 'innovative, market-based financing mechanisms', such as 'the voluntary carbon market', while downplaying the emissions-trading fraud, corruption, speculation and collapse, now rife across the world.

Regrettably, ASSAf believes in a few other 'false solutions' to the climate crisis, such as biofuels (Durban is a sugar cane centre) and the dangerous co-incineration of tyres in cement kilns. And in another disturbing development, ASSAf's emphasis on residents' behavioural change risks a blame-the-victim mentality, for example, justifying the denial of flush toilets for poor people, to avoid increased electricity use at the sewage works. Remarked Diab at the ASSAf launch: 'We must encourage people to stop using their cars and start using public transport', yet ASSAf was silent about how city officials let a crony-capitalist firm, Remant Alton, privatise and wreck the municipal bus system.

Not a total write-off, ASSAf's report at least encourages Durban to 'produce local, buy local' at a time of inane currency-induced trading patterns, that have little to do with rational comparative advantages between competing economies. The report condemns suburban sprawl and much post-apartheid planning, while endorsing the 'polluter pays' principle, which, if ever implemented, would radically improve the city's environment. All obvious enough, but what hope for implementation, given Durban rulers' pro-pollution bias? 'Climate smart,' according to Roberts, means a city's 'low-carbon, green economy provides opportunities for both climate change mitigation and adaptation, and fosters a new form of urban development that ensures ecological integrity and human well-being.' Precisely. But if Diab is correct that 'poor public awareness' is a major barrier to addressing the most serious crisis humanity has ever faced, ASSAf scientists now contribute to that very problem with their bland, blind greenwashing of climate-dumb Durban.

Durban's contestations

From such contradictions arise intense contestations, which usually turn violent from the top down. Police are regularly let loose on communities that attempt to stage non-violent marches. Shack dwellers, market traders, homeless children, organised labour and residents' organisations are treated with contempt by city officials. In the three years prior to the COP 17, to take an arbitrary date, protests were passionate, especially against polluting industries and a neoliberal, repressive municipality that regularly denies demonstrators their rights to protest. Consider some of these in chronological order.

In January 2009, the Qadi people at Inanda Dam revolted over their uncompensated land claim and were denied permission to protest at the Dusi Canoe Marathon. DUT staff and students went on strike over salary and fee grievances. In February, DUT was joined by UKZN students. There were municipal service protests in Ntuzuma, and protests against police violence in Umlazi. In March, DUT students protested again, as did bus commuters and the South Durban Community Environmental Alliance (SDCEA), angry at the lack of state regulation of the oil industry. In April, community activists tackled inadequate housing in KwaMashu and Lamontville, while traders at the Warwick Junction Early Morning Market put the city on notice that a shopping mall would not displace their century-old institution. In May, there were protests against forced removals and inadequate housing in Siyanda, Marianhill and Lamontville, while the SDCEA tackled Shell Oil, and Warwick traders continued their anti-displacement protests. In June, there were demonstrations by victims of electricity disconnections and shack fires at Kennedy Road, and of substandard housing in Lindelani, as well as conflicts over conditions in the Umlazi hostels. The local chapter of Zimbabwe's Movement for Democratic Change protested at Home Affairs and won concessions.

Between June and July 2009, the Early Morning Market protests grew more intense. Public sector doctors were on strike, and the National Union of Mineworkers members fought for higher wages from Moses Mabhida Stadium building contractors, as did Durban bus drivers, who demanded the retention of routes that failed privatisation threatened to close. Residents joined the bus drivers protesting Durban transport in July, and the South African Municipal Workers Union (SAMWU) struck the municipality. Hungry, the South African Unemployed People's Movement protested in two major downtown food stores, eating while sitting-in, suffering more than 100 arrests but making a visceral point. In August, Warwick traders continued their struggle against eviction. Other public sector workers on strike included the Communications Workers Union against the Post Office, Telkom workers versus Telkom, and social workers against the provincial government.

There were protests by Lamontville residents against poor housing, while nearby, Clairwood and Bluff residents protested truckers. In September, Durban's tow truck drivers protested Johannesburg competitors in the context of weak regulation. In October, the Abahlali baseMjondolo movement (evicted from Kennedy Road the previous month in an attack) won a court victory over the provincial Slums Act. In Chatsworth, communities protested the toxic Bulbul Drive dump, eventually leading to prosecution of the corporation running the landfill, and ultimately winning closure in 2010. Mangosuthu University of Technology (MUT) students protested fees, and ratepayers associations complained about electricity price increases. In November, residents of Wentworth and other South Durban communities marched in central Durban against low-quality, high-priced service delivery, while residents of Hammarsdale and Mayville also demonstrated for better services. Mariannridge activists demanded housing, and Phoenix residents were up in arms over poor treatment at the Mahatma Gandhi Memorial Hospital. UKZN workers began regular protests against the increasingly common and exploitative practice of labour broking.

In 2010, protests were even more intense, given labour unrest mainly over wages and outsourcing. In January, MUT and DUT students demanded lower tuition fees.

South Durban residents protested the municipal regulation of truckers, while Umlazi hostel residents demanded urgent upgrading. In February, the students continued demonstrating. In Verulam, traders attacked the municipality over rental policy. In the south, SDCEA began national protests against the US$3.75 billion World Bank loan to Eskom. Inner-city flat dwellers and bus commuters also protested municipal policies. In March, there were outsourced labour protests against UKZN security and transport operations, a food-price protest at the DUT, as well as a protest by Abahlali baseMjondolo against state repression. In April, these were joined by UKZN students demanding better accommodation and an end to financial-based exclusions, while SAMWU embarked on a strike, and the South African Communist Party marched against rampant state corruption. In May, the South African Transport and Allied Workers Union had a successful wage strike that paralysed the harbour, leaving hundreds of boats at sea unable to offload. In June, there were two protests against the World Cup: by Stallion Security companies over pay at the first match held in the R3. billion (around US$450 million) Moses Mabhida Stadium, and then by Durban communities against the 'white elephant' stadium. In July, communities gathered to protest xenophobia (and its causes in state and business policies), and fisherfolk were arrested after being denied access to the main piers. In August, the massive public sector strike began, affecting specifically health facilities and schools, as well as Home Affairs. There were also protests against sweatshops in the vicinity of Durban, and in Umlazi there were service delivery protests. In September, UKZN student accommodation protests were followed by South Durban environment demonstrations in November and December (the latter as part of the '1 000 Cancúns' called by Via Campesina).

Student protests over fees and exclusions began this year in Durban, compelling a DUT shutdown in early February (repeated between July and August) and ongoing protests at several UKZN campuses. By late March, police action against students with tear gas and rubber bullets led to scores of injuries, and the escalation continued for another week. In April, water cannons and rubber bullets were also used against MUT students in Umlazi. From March to June, numerous mass protests were registered in relation to the May municipal elections, including that of the Inkatha Freedom Party (IFP) against the breakaway National Freedom Party (NFP), and the ANC, as well as internecine ANC battles.

In June, KwaDukuza was the site of major protests by, and repression of, commuters angered by rising transport price hikes. There were 155 arrests, and police killed one protester. In July, hundreds of fisherfolk protested against municipal and provincial restrictions on fishing, and thousands of metalworkers and Congress of South African

Trade Union activists marched against police brutality and workplace exploitation in Pinetown and central Durban. Between July and August, protests over municipal disconnections of electricity occurred at Sea Cow Lake, near the N2 highway (leaving one municipal subcontractor dead), at Kennedy Road in Clare Estate (with police shootings), and in Chatsworth where Westcliff flat dwellers succeeded in reversing a downgrading of electricity from 60 to 40 amps.

Although aside from the Bulbul Drive dump closure, the only genuine victories (aside from wage struggles, won in late 2008 by Abahlali baseMjondolo, and mid-2009 by the Warwick Early Morning Market activists) were defensive in nature; the vast majority of protests mentioned above are inspiring reflections of resistance. And there were probably scores more expressions of protest across the city, during this period, not covered by journalists, and hence by the Centre for Civil Society Social Protest Observatory. (Durban's two main English papers, *The Mercury* and *Daily News,* have been victims of cutbacks in coverage, as its Independent Newspapers owners reduced coverage and sent profits abroad.)

But it must also be acknowledged that protest sometimes carries the highest personal price. Durban can be an extremely dangerous place for activists, dating back decades. Earlier high-profile assassinations of political philosopher, Rick Turner (1978), Griffiths Mxenge (1981) and Victoria Mxenge (1985) were carried out by apartheid hit squads. The 1980s and 1990s witnessed thousands of murders of both ANC-aligned and IFP activists, though the preponderance of evidence implicates the PW Botha and FW de Klerk regimes' collaboration with IFP, including arming its leadership. But somehow, even after the ANC-IFP rivalry was won by the former at the ballot box, and after the Zulu nationalists faded badly as a rural political party with a few strongholds among Durban hostel-dwellers, the method of settling local political scores with bullets continued.

On July 11, 2011, for example, Durban's leading ruling party official, Sbu Sibiya, was assassinated at his home, four months after ANC councillor, Wiseman Mshibe, was shot dead. In early August, 2011, Gundu Makhanya, a Chatsworth leader of the NFP was killed, with allegations that the IFP was responsible. Across the province, several other NFP leaders were killed in the weeks before and after. The other leading party with African township support, the Congress of the People, suffered extreme harassment by the ANC just before the 2009 national elections. In early 2009 in Umlazi township, a hail of 50 AK-47 bullets killed traditional leader, Inkosi Mbongeleni Zondi (grandson of Bhambatha Zondi, the Zulu leader who led a rebellion against Natal colonial officials in 1906). A month later in the suburb of New Germany, the local leader of the South African National Civic Organisation, Jimmy Mtolo, was shot on a Saturday morning by an assassin who came into his office, ostensibly seeking help with housing. His daughter, ANC Member of Parliament, Ntombikayise Sibhidla, complained about police failure to find the murderer. A few weeks later, in April 2009, the next assassination victim was SDCEA leader, Ahmed Osman, gunned down on his stoep in cold blood on a warm late-summer evening, in the midst of fighting companies responsible for toxic waste releases in Clairwood. In 2008, SDCEA's Des D'Sa was nearly killed in a firebombing of his flat. In 2007, ward councillor and former SDCEA member, Rajah Naidoo, was the victim of a similar hit.

In August 2008, demonstrations in the centre of town in front of the ICC led to the death of 22-year-old University of South Africa political science student, Mthoko Nkwanyana, who was protesting high tuition fees alongside 400 others; police used tear gas so aggressively that he died in the skirmish. This was reminiscent of a similar protest at UKZN's Westville campus in 2001, when Michael Makhabane was killed by security guards during a peaceful protest of more than 500 students, again against high fees. Another youth, Marcel King, was killed in June 2004 in Phoenix township, not far from Gandhi's settlement, shot between the eyes by security guards hired by the Durban municipality to disconnect illegal electricity hook-ups. According to his brother, Jonathan, 'My mother was involved in a confrontation with a security guard who had just hit her, because she had tried to climb on to the back of his van. Marcel tried to pull her away but the guard cocked his gun and began firing.'

In July 2007, civil society activist, Sajida Khan, died because Durban municipal toxins, floating across Clare Road from Bisasar Road dump, gave her two bouts of cancer. The most famous post-apartheid death cannot be attributed to government, opposition parties, business or gangs, though: Activist, Gugu Dlamini, was killed in December 2008 by Umlazi residents, who stigmatised her AIDS education work and stoned her to death. Activists mourn them all, including two giant leaders of struggle and independent progressive politics, who died peacefully in late 2009 and early 2010, respectively, Dennis Brutus and Fatima Meer.

A contested COP during a capitalist crisis

The contributions in the pages that follow explore environment, segregation, capital accumulation and power. We begin with Patrick Bond and Ashwin Desai on the spatial squeeze Durban is witnessing. Writ large it's a case study of environmental injustice, insofar as a great many low-income people are suffering displacement from homes, streets and even fishing piers in the area adjoining the ICC. Desai and Goolam Vahed then describe the trajectory of the nearest residential neighbourhood to the ICC, and thereafter, Desai covers emblematic street-level aspects of the cityscape in an essay on sightseeing Durban, and Bond reviews the myriad eco-social struggles over the city's hydrological system (which will be one of the first victims of climate change). Next, from the notorious petro-chemical hub of Wentworth, a rigorous community-based analysis conjoins environmental justice struggles across the metropolis. The SDCEA envisages an anti-capitalist alliance of the various eco-social campaigns that make Durban so compelling, notwithstanding difficulties in stitching together the component parts so far.

Subsequent contributions take the premise that Durban's COP 17 is mainly about the financing of climate management, publicly sourced in the GCF and privately through carbon markets, with some mix envisaged by global elites during a time of severe and worsening fiscal austerity. Del Weston suggests that the host government will be incapable of promoting emissions cuts globally, given its own fossil fuel addiction. Khadija Sharife provides details about the Durban experience of the CDM technique, and debates the timber offsets in Africa. Larry Lohmann considers the deeper contradictions of carbon markets. Oscar Reyes describes foibles in the new sectoral climate investment strategies. Finally, Yash Tandon warns that the GCF co-chaired by South African Planning Minister Trevor Manuel, the continent's leading neoliberal, could devastate Africa.

In sum, there is no question that Kyoto Protocol proponents, including the South African government, are failing to generate the support required for a new round of post-2012, legally binding commitments. So the outcome of the Durban COP is already assured: *failure*. Moreover, if the climate-financing quandaries play out as expected, then the most durable accomplishment of Durban will be a deepened commodification strategy. At the same time, sophisticated financiers understand that the ongoing world capitalist crisis (in mid-2011, reflected in the Standard & Poor's downgrading of Washington's credit rating, rising default fears on the European periphery, and new problems with China's Keynesian financing of infrastructure, overlaid upon stressed-capitalism catastrophes such as nuclear reactors, oil spills, flooding, and all manner of social upheavals) throws into question the capacity of markets to assist in climate management.

To illustrate, on August 10, 2011, the *Wall Street Journal* asked New York University business professor, Nouriel Roubini, 'What can government and what can businesses do to get the economy going again, or is it just sit and wait and gut it out?' His answer was, as usual, insightful:

> 'Businesses are not doing anything. They claim they're doing cutbacks because there's excess capacity and not adding workers because there's not enough final demand, but there's a paradox, a Catch-22. If you're not hiring workers, there's not enough labo[u]r income, enough consumer confidence, enough consumption, not enough final demand... In the last two or three years, we've actually had a worsening because we've had a massive redistribution of income from labo[u]r to capital, from wages to profits, and the inequality of income has increased. And the marginal propensity to spend of a household is greater than the marginal propensity of a firm because they have a greater propensity to save, that is firms compared to households. So the redistribution of income and wealth makes the problem of inadequate aggregate demand even worse... Karl Marx had it right. At some point, capitalism can destroy itself. You cannot keep on shifting income from labo[u]r to capital without having an excess capacity and a lack of aggregate demand. That's what has happened. We thought that markets worked. They're not working.' (Heaven 2011; Wall Street Journal 2011)

Nor can they work in the reorganisation of central Durban or in climate management for reasons we will observe in the pages that follow. First, limits of market-centric strategies are being reached for Durban's neoliberal municipal state, notwithstanding massive national commitments to sports tourism, international conferencing, the petrochemicals industry, Durban's municipal services policy and export-led growth (all of which have adversely affected the socio-ecological balance in central Durban). Attempts by this local state to channel investment through unprecedented capital grants and operating subsidies, are not only creating white elephants (even municipal-controlled retail and restaurant space across the renovated beach front has been unlettable since the World Cup), but also more pollution, including accelerated climate change in air travel, tourism, and shipping, and more gentrification-style squeezing of poor people, street children, immigrants, informal traders, and fisherfolk. Limits are appearing because capital isn't following the state's expensive pushing and pulling, and because of resistance from below.

Secondly, the limits of the market for solving climate crises, via carbon trading, are now clearly evident, what with the 2011 collapse of the European Emissions Trading

Scheme through hacking and credit-theft in January, thanks to a general lack of confidence in subsequent months, and the 2010 demise of Chicago's carbon exchange. As for potential new markets, the biggest is California, but the California Air Resources Board's announced emissions trading system was hit hard by a series of 2011 legal challenges from progressive environmentalists, and may be completely redesigned to avoid trading. Australian and Korean prototypes are being delayed. Although offsets, Reducing Emissions through Deforestation and forest Degradation (REDD) and Clean Development Mechanism (CDM) gimmicks continue to spread, the underlying basis for these is shrivelling, especially if in Durban the Kyoto Protocol's emissions cuts targets are not formally extended past 2012, and Kyoto's CDMs are ultimately shelved. Of course, the worst-case outcome is also worth contemplating, because it is apparently the basis for a growing Northern consensus: Kyoto's binding emissions cuts fail to be respected and extended into a post-2012 era, but the Kyoto carbon trading mechanisms are preserved.

This trial of carbon capital, and the demise of the Kyoto Protocol, together leave a huge opening for climate justice advocates, as it becomes clear that the elites have nothing substantive to offer. The forces arrayed in Durban will generally support the most progressive framing of climate politics possible, and so the stage is set for a dead inside space, and a very live time outside the UNFCCC.

References

Association for Science in South Africa. 2006. *Towards a low carbon city.* Pretoria. August, http://www.assaf.org.za/2011/08/durban-on-a-pathway-towards-a-low-carbon-city/.

Bond, P. 2006. *Talk left walk right.* Pietermaritzburg: University of KwaZulu-Natal Press.

Bond, P., R. Dada, and G. Erion. 2009. *Climate change, carbon trading and civil society.* Pietermaritzburg: University of KwaZulu-Natal Press.

de Boer, H. 2007. Embarrassing meaning of eThekwini revealed. *Daily News.* August 22.

Desai, A. 2002. *We are the poors.* New York: Monthly Review Press.

Gandhi, M.K. 1928. *Satyagraha in South Africa.* Navajivan: Ahmedabad.

Heaven, P. 2011. Marx was right. *Financial Post.* August 12, http://business.financialpost.com/2011/08/12/roubini-says-more-than-50-chance-of-global-recession/.

Kell, D. and C. Nizzo. 2011. *Dear Mandela*, http://www.dearmandela.com/.

Vedantam, S. 2005. Kyoto credits system aids the rich, some say. *The Washington Post.* March 12.

Wall Street Journal. 2011. Interview with Nouriel Roubini. August 12, http://online.wsj.com/article/68EE8F89-EC24-42F8-9B9D-47B510E473B0.html.

PART ONE:

DURBAN'S POLITICAL ECOLOGY

Chapter 1

Prefigurative Political Ecology and Socio-Environmental Injustice in Central Durban

Patrick Bond and Ashwin Desai

I stand and gaze and feel—and marvel! Is
This then the great city that has planted
Despair in me? What contrasts jolt in this
Strange Hive: souls kind and hard; pure Good; great Sins!
This Hope or Mockery, Lord? Or Joy or Pain?
For here beneath my eyes lie wonder scenes
That should ring Joy but only fling me Pain!
Herbert Dhlomo, 1940s
(from Berea Ridge above central Durban)

Sturdy British businessmen
made this town (and "Coolies"!)
Light-festoons along the beach
dribble away the nutriment
of glaucous hunger-swollen urchins;
proud men display perverse inverted pride
as carrier-beasts for lording colonialness;
the stairs are occupied by ancient odors
of curry, hospitality and insecurity,
and the image of incited rampage

is cherished like a rusted hunting knife
amid the bustle of rapacity,
uncertain liberals and pink gins.
Dennis Brutus, 1960s

Two of the city's poets of yesteryear, Herbert Dhlomo and Dennis Brutus, would undoubtedly look at Durban from the western heights today, and pass harsh judgment on a political transition that replaced racial barriers with class barriers. And this on a sport-tourism paradise, replete with South Africa's most extreme eco-social contradictions, and on redevelopment processes that further squeeze and pollute the poorer residents. Yet no one will deny that the city of Durban has prepared an extraordinary setting to host the 17th Conference of the Parties (COP 17) to the Kyoto Protocol, from November 28 to December 9, 2011. The International Convention Centre Durban (ICC Durban) and the adjacent Durban Exhibition Centre (DEC) together constitute the largest conference facility in Africa. Within a few blocks, hotels serving middle- and upper-class clients adjoin the most democratic public space in South Africa – Durban's beachfront – which was rehabilitated just before the soccer World Cup 2010.

New highway extensions into the city, and a new airport and beachfront drive, give visitors a particularly luxurious entrance to a South African city. The view, as you enter the sporting precinct from the wealthy neighbourhoods of Durban North, is simply spectacular with golf courses on both sides giving way to two world-class stadiums on your right, and exceptional beaches on your left to the east. Durban's official plans for hosting the COP 17 allow delegates to begin accreditation next to the new Moses Mabhida Stadium, followed by security checks near the main casino, and a beach party at the site of the celebrated 2010 World Cup Fan Fest. To the south is the Workshop Shopping Centre and a still-glamorous Durban City Hall, surrounded by palm trees, but incongruously designed in 1910 using Northern Ireland's Belfast City Hall as a model.

This site reveals a great deal about both Durban politics and the potential for disruptions to the desired outcome of the COP 17. As constituted by the UN Framework Convention on Climate Change (UNFCCC), that outcome is a straight neoliberal trajectory from Copenhagen's COP 15, whose 2009 Accord between the US, Brazil, South Africa, India and China 'blew up the United Nations', as Bill McKibbon put it (Bond 2010), to Cancún's COP 16 revival of carbon market fantasies (Bond 2011), to Durban where climate finance deal-making appears as the only substantive augmentation to global climate governance. By all accounts, Durban will spell the end of the Kyoto Protocol's binding commitments and differentiated responsibilities. The name of the city will be, for climate science and activism, akin to Oslo's reputation in the Middle East: betrayal, incompetence, elite failure.

What of socio-political resistance? Even if there is no disruptive blockade, the moment is ripe for comparison to the powerful joining of radical activists, labour, social movements and environmentalists, who showed that the obscure World Trade Organization's Millennium Round in Seattle, in 1999, could give enormous leverage, not only to highlight single issues, but to thread them together with like-minded critics of a universal opponent. Will climate change – and a growing popular critique of the 'Conference of Polluters (COP)' – potentially offer a similar opportunity for connecting dots, finding convergences and forging unity? Will the area around the negotiators create a *laager* in which the paranoid political economy of global conferencing aligns with Durban's distorted political ecology of exclusion and environmental destruction? Or will the strands of dissent in the area help tens of thousands of incoming protesters to find unity, across traditional divides of community, labour and environment?

To get a sense of the possibilities in the immediate vicinity of the ICC, we begin by noting contradictions embodied in the sporting precinct, ready for further contestation given the legacy of hosting the World Cup and the envisaged bid for a future (2024) Olympics. We then move southwest, exploring Albert Park through the eyes of oppressed immigrants, before considering the Indian quarter and the critical site of Warwick Junction, where, in 2009, informal traders beat back an attack by the city. From there, due east towards the beaches, a less encouraging spatial struggle between fisherfolk and municipal officials is revealed. Together, these struggles clarify future lines of engagement, as the climate crisis intensifies Durban's eco-social contradictions.

Sport Tourism as Climate Injustice

Just north of the ICC, the world-class Moses Mabhida Stadium – named after an anti-apartheid Communist Party organiser, who was a mentor to South African President Jacob Zuma – cost R3.4 billion (around US$450 million), overrun from an original R1.8 billion budget. The 70 000 seater Moses Mabhida Stadium is delightful to view, as long as we keep out of sight and mind the city's vast backlogs of unmet needs for housing, water and sanitation, electricity, clinics, schools and roads – all of which is punctuated by the absurd cost escalation.

In 2009, while redirecting municipal resources to the Moses Mabhida Stadium, top Durban officials decided to reduce the annual target of 16 000 new low-cost houses, built to just 9 500, as against a (conservatively-estimated) backlog of 234 000 units (Academy of Science of South Africa 2011:41).

Even harder to keep from view is the next-door neighbour, Kings Park Stadium (previously known as the ABSA Stadium) which is home to the Sharks rugby team. This stadium seats 52 000 and could easily have been extended for the World Cup, since only a year earlier it had hosted a Confederations Cup semi-final match.

As is typical for cities hosting large international sporting events, Durban experienced a hefty World Cup hangover, mainly due to the ongoing losses from Moses Mabhida Stadium. The World Cup Local Organising Committee (LOC) leader, Danny Jordaan, apologised to the country in mid-2011 because, as he put it, 'not enough thought had gone into (the) planning of the stadiums to ensure [that] they would be financially viable'. Durban City Manager, Mike Sutcliffe, disagreed: 'I am happy. We made the right decisions.' But in contrast to the money-earning Kings Park Stadium, Moses Mabhida Stadium's annual R80 million (around US$12 million) in operating costs were only partially offset by income of just R56 million in the highest-use year of 2010–2011 (Dardagan 2011). Sutcliffe's efforts to move the Sharks across the road failed repeatedly, so long-term municipal subsidisation is required for Moses Mabhida Stadium, as is the case with the ICC and Point's uShaka Marine World, the city's other two large white elephants.

The sharpest protest against the stadium's operation was by several hundred security workers on June 13, 2010. Immediately after the Germany-Australia game, the workers demanded payment of a promised bonus, since they only received R200 (around US$30) for 12 hours' work. This paltry pay is a reflection of the outsourcing and super-exploitation now typical in the often-dangerous security sector. Police used teargas and rubber bullets to control the striking workers, 300 of whom then joined more than 1 000 other community- and social-movement activists in a June 16 protest march to City Hall to demand a 'World Cup for all'.

On July 3, several hundred people rallied (again) at City Hall against what might be regarded as the greatest failing of the local subaltern classes: xenophobic tendencies. The inner-city hosts a large immigrant population, and just as in mid-2008, the period immediately after the World Cup, was rife with attacks on Southern African immigrant workers by South African workers. The attacks were based on the widespread perception that immigrants take jobs, lower the prevailing wage rates, overcrowd scarce accommodation in expensive inner-city housing markets, and bring excess competition to micro-retail township markets. Although undeniable, these factors then require even more vigorous community activism to address the structural roots of xenophobia, instead of treating the cancer merely as a superficial blemish (Amisi et al. 2011).

While these actions appeared futile in contrast, the most successful protest – explicitly against the World Cup – was carried out in mid-2009 by hundreds of Durban's informal traders facing displacement from the century-old Early Morning Market, which city planners wanted to raze in order to erect a nearby shopping mall in time for the World Cup. However, sustained resistance over a year-long period, including a pitched battle with police in July 2009, saved the market. In another form of protest, culture-jamming, Durban's leading rapper, Iain 'Ewok' Robinson (2010), contributed his thoughts on the World Cup with the track "Shame on the Beautiful Game".

The main winners in the Mabhida debacle were large corporations and politically-connected black 'tenderpreneurs' – who win state tenders thanks to affirmative action, if they are connected to established white firms – especially in the construction sector. Durban journalist Sam Sole (2010) posed the question of whether Mabhida's legacy is an 'arch of hope' or a 'yoke of debt', in a hard-hitting investigation of 2010 World Cup crony-capitalism. According to Sole, the main beneficiaries were Ibhola Lethu (Our Football Team), the lead consortium for the stadium, which included Craig Simmer[1]; the old Broederbonder engineering firm, BKS, and the construction firm, Group 5/ WBHO (part of former state prosecutor Bulelani Ngcuka's Mvelaphanda Group, whose chairman at the time was the current Minister of Human Settlements, Tokyo Sexwale). Durban tycoon Vivian Reddy's Edison Power also scored a major electricity contract in the deal.

Another questionable project was the new R8.2 billion (around US$1.2 billion) King Shaka International Airport (KSIA), given that Durban International Airport (DIA) in South Durban had excess capacity until 2017, and that there is no public transport to KSIA, which is twice the distance from central Durban. The premature closure of DIA proved the single biggest embarrassment for the LOC leadership and the Airports Company of South Africa, when KSIA could not cope with VIP private jets that jammed the airport on July 7, the day of the semi-final between Spain and Germany. More than 1 000 furious soccer fans in eight planes had to be re-routed back to Johannesburg and Cape Town because the VIPs blocked boarding gates, refusing to park their jets away from the runway or fly to other nearby (smaller) airports.

The contrast between Durban's extreme sport tourism extravagances, and the worsening living conditions of the vast majority of its citizens, is acute. Leading up to and during the World Cup, Durban's underclass engaged in creative and poignant actions to expose their harsh living conditions. One such action was the 2008 'eat-in' by hungry, unemployed people, who on two occasions simply sat and ate food within a local supermarket without paying, resulting in 200 arrests. The ongoing attempts by street children to avoid arrest also highlighted the discrepancy.

Generational forced removals

Given that climate change will have an adverse effect along generational lines, it is particularly disturbing that some of central Durban's most excluded people are children – especially those whose families have been disrupted by HIV/AIDS (and the initial refusal by the former Mbeki government to provide antiretroviral medicines, leading to an estimated 330 000 unnecessary deaths, as estimated by a team from the Harvard School of Public Health). If past practices continue, street children will be removed from the area around the ICC for COP 17. Although the municipality was pressured to adopt a more humane approach in preparation for the 2010 World Cup, its intentions were obvious for several years prior. In 2005, for example, the City Manager made this revealing official statement:

> Screaming headlines claiming that eThekwini [Durban municipal government] is rounding up street children and ushering them out of the public's eyes to avoid them being seen by tourists, provide evidence once again of the shoddy state of some of our media and their lack of transformation… We can understand outrage if we were rounding children up and dumping them in jails or the outskirts of the city (Sutcliffe 2005).

All indications in subsequent years confirmed that dumping children outside Durban *was* common, and in 2009, Sutcliffe, quoted in Durban's daily newspaper *The Mercury*, conceded as much:

> When the heads of state are going to appear at conferences, the first thing that security does is move the street children out of the way so they can't be seen. Four years ago we used to deal with it that way, but we are not going to deal with it like that anymore (Comins 2009a).

But as the World Cup approached, matters degenerated, for as the *Sunday Times* reported in early 2010:

> Street children have accused the Durban metro police of removing them from the city. This week, youngsters from nine to 19, told the *Sunday Times* how metro policemen rounded them up daily and dumped them on the city's outskirts, often on busy freeways or at unregistered shelters. …[l]ocal NGO Umthombo said the police's methods were inhumane. The organisation said the welfare of the children was more important than Durban's image. CEO Tom Hewitt said: "Street children are not a safety and security issue, but a social development issue. Police put children in the back of trucks with adults and gang members. It exposes the kids to more trauma." He said removing children for the World Cup was not about child protection but about cleaning up the streets (Laganparsad 2010).

Hewitt (2010) elaborated on the problem on Umthombo's website:

> Round-ups have often happened in Durban before major international conferences or sports events. Now, with the World Cup in mind, it appears metro police have stepped up their efforts. The children allege that the police sometimes attempt to take them back to their homes, but end up dumping them on the roadside. Seventeen-year-old Wendy Ndlovu said she and two friends were recently left in Amanzimtoti, 30 kilometres away. "[The police] dropped us near the garage. I asked the people where town was and we started walking. We walked all night and we slept in the bushes," said Ndlovu. "[The police] say we can't be here [in the city] for the World Cup and that they will take us to a nice place. But they end up leaving us anywhere or in these scary houses."

So much for repeated commitments by top city officials to treat street children with dignity. With respect to intergenerational justice, street children are among Durban's main victims of economic injustice, the healthcare crisis, and in the future, climate change.

Xenophobia in Durban: two deaths in Albert Park

The other central-Durban residents who are the targets of similar municipal – and social – disregard, are refugees from areas such as Southern Matabeleland in Zimbabwe, who (due to linguistic similarities with the local Zulu language) are the majority immigrant group in the city. In the future, they will be among the continent's hardest-hit climate change victims, given their vulnerability to drought. In the zone southwest of the ICC, which encompasses the decaying inner-city, medium-rise flats around Albert Park, local working-class men have repeatedly expressed their hostility to these and other refugees. In January 2009, this hostility took a fatal turn.

An outbreak of xenophobic violence occurred against African immigrants, and the main attack was allegedly led by the African National Congress (ANC) city councillor from Albert Park, Vusi Khoza. It resulted in the deaths of two people who were thrown from the Venture Africa building's sixth floor. The city authorities' response (permitting Khoza to remain a councillor) and the ongoing court case shed light on how difficult it is for immigrants to find a secure place in central Durban. Southern Africans, especially from Zimbabwe, Mozambique, the Democratic Republic of the Congo (DRC), Tanzania, and Malawi, as well as parts of the Horn (Ethiopia and Somalia) and West Africa (Nigeria) have witnessed an extraordinary rise in hostility from South Africans over the past decade.

This harsh reality shatters the dream enticing African immigrants to South Africa – a dream that is, ironically, often encouraged by highly romantic descriptions of what awaits those who come. Here, for the benefit of tourism marketing, Zweli Mkhize (later the provincial Premier) sings the praises of the

> 'cosmopolitan appearance of the population, the multitudes of cultures [and] tolerance of the diversity of cultures. …availability of a diverse kind of goods, particularly when we talk about the arts and crafts, that are very specific in terms of identity – that these are what you will find in an African setting. And it's increasingly getting very cosmopolitan in terms of African states, that you will find actually showing their influence around. So it makes you very much… able to say once you are here, you are in a friendly environment' (quoted in Bass 2009:253–254).

Two senior city officials, Sutcliffe (the City Manager since 2002, although slated for replacement in late 2011) and his assistant responsible for the ICC, Julie-May Ellingson, espouse Durban's official determination to emphasise an African City identity, with the statement that 'Our theatre is Africa, South Africa is the stage, and we (Durban) are one of the main characters' (cited in Bass 2009:250). Bass (2009:250) quotes Sutcliffe's claim to 'creating an inclusive, well-governed, caring and democratic city'. But the experiences of African foreigners on the streets of the city of Durban – which have been marked by the most vicious form of *othering* – expose the hypocrisy of these sentiments.

Across the country, the May 2008 xenophobic attacks resulted in the deaths of just over 60 people. According to official reports, some 342 shops were looted, 213 gutted, and 1 384 people arrested (Minister of Safety and Security Charles Nqakula, quoted in Crush et al. 2008:11). The attacks were met with shock and horror across South Africa and further afield. How could this be happening in a country with an international reputation for reconciliation, and whose people were dubbed the 'rainbow nation of God' in recognition of their seemingly miraculous ability to overcome a three-century-old racial division and its associated oppression? It is also a country that is widely acknowledged for its founding constitution, and emphasis on human rights supported by a relatively well-funded domestic Human Rights Commission.

Speaking on behalf of the government, Minister of Intelligence Ronnie Kasrils blamed a mythologised 'Third Force' of political saboteurs and other right-wing elements and criminals. There was little acknowledgement that xenophobic violence had a longer history in post-apartheid South Africa. When warned by the African Peer Review Mechanism team in December 2007, that 'xenophobia against other Africans is currently on the rise and must be nipped in the bud', Thabo Mbeki, the then South African President, replied that this was 'simply not true' (Johwa 2008). But from the birth of the new democracy, the danger signals were there. As early as December 1994, African foreigners were attacked by local residents in Alexandra township, next to Johannesburg's financial centre, where many witnessed their shacks and houses destroyed, while others were forcibly marched to the police station. The mob mobilisation was named 'Operation Buyelekhaya' (go back home). According to Jonathan Crush et al. (2008:21),

> attacks on foreign nationals escalated in their brazenness and brutality. In a spate of attacks in 2007, over 100 Somalis were killed and Somali businesses and properties were looted and torched. Certainly there were plenty of danger signs. Government ministers should not have been surprised in May 2008.

Comprehensive surveys pointed to strong and growing xenophobic attitudes (Mattes et al. 1999). While xenophobic sentiments against African immigrants were pervasive and had trespassed into the organs of the state and translated into violence, the May 2008 attacks were different. The previous violence was limited to one episode in one locale, while the 2008 attacks spread across the country, putting tens of thousands of people on the move. At least 15 000 Mozambicans left the country in a convoy of buses, while thousands of other African refugees sought sanctuary in camps and makeshift shelters.

Even those who fled unspeakable horror in their home countries – which made returning home inconceivable – were targeted. In the eastern Democratic Republic of the Congo, an estimated 5.4 million people were killed between 1998 and 2008 (Reuters 2008), in part because of the multinational corporate search for coltan, which is used ubiquitously in cellphones and other common electronics, and is extracted with the assistance of warlords. The international gold producer of South African origin, AngloGold Ashanti, had operations in a 2 000 square kilometre tract of land around Mongbwalu, after dubiously acquiring exploration rights there in 1996, during the reign of dictator Mobutu Sese Seko (Deibert 2008).

In 2005, Human Rights Watch revealed AngloGold Ashanti's 2005 payments to warlords of the Front for National Integration (FIN), the leader of which was prosecuted in The Hague's International Criminal Court on war crimes charges. 'Our central purpose is to find and mine gold profitably', explained its then chief executive, Bobby Godsell, adding that '[m]istakes will be made' (Godsell 2005). The South African state strongly supported Johannesburg capital to enter the area during the early 2000s (Bond 2002). The blowback of regional sub-imperial power, for the South African state and capital, was a flood of terrorised refugees, generated as a result of Pretoria's foreign policy on behalf of Johannesburg mining houses' hunger for minerals extraction.

Durban witnessed this blowback when a group of 47 refugees from the eastern DRC, who needed shelter in the aftermath of the May 2008 attacks, ended up at Albert Park, a run-down field at the south-western side of the central business district, surrounded by dilapidated flats. In the park, they were constantly harassed by police. The refugees begged the city to turn the place into an official camp. Local NGOs (some of them in the pay of the United Nations High Commissioner for Refugees UNHCR) tried to convince the group to disperse. City authorities used particularly cruel and insidious language in discussing the issue. For Sutcliffe, the issue was very simple:

> 'The families have a choice of either going back to their countries, or to places in the community they were living in before the July problems. The municipality cannot suddenly prioritise their housing needs when we already have 200 000 [local] people with housing needs in the city itself' (Mdletshe 2008).

The refugees, in fact, had not demanded formal housing; all they wanted was to have a sanctuary in Albert Park. It was an insinuation that could only isolate the refugees further and act as a cover for the city's inability to make good on its promises of housing the poor. Sutcliffe's rationale for a speedy removal had another basis: 'I have instructed the metro police to remove people from the park because the surrounding community has complained about crime' (Magwaza 2008). The fact that none of the refugees was fingered for any crime did not seem to matter. But there was another factor: the 2010 World Cup, less than two years away.

In early November 2008, the city arrived in the form of Durban Solid Waste and a strong contingent of metro police. This time, the rationale was that the park needed to be cleaned up. One of the 47, Aziza Wilondja, a mother of six who had stayed in the park for four months, spoke of how their clothes and identity documents were confiscated:

> *'The police took our things and put them in the garbage vehicle. They brought down our tents and threatened to beat us.'*

Initially collectively gathered in a bid to hang onto the bare rudiments of their lives, the refugees were now forced to leave Albert Park and embark on separate, individual journeys. For the City Manager, they were a problem that had to be made to disappear, warning that, while the support of NGOs and the public was to be commended:

> 'We also have to take care that we do not perpetuate the situation longer than necessary. A growing refugee problem is something we must try and avoid at all costs. (Bond, Hinely, and Meth' 2008).

Following direct orders from Sutcliffe to remove the Congolese,[2] the police officer in charge, Constable Kwesi Matenjwa, spoke on tape to the University of KwaZulu-Natal's Centre for Civil Society about the forced removal and confiscation of goods (including official refugee papers), in revealing detail about the squeeze associated with the World Cup:

2010 is going to be here, so the people from the so-called other countries, when they come to this country, they must have this image that South Africa, the city of Durban is clean, that there are no vagrant people, there are no traders in the streets (Bond, Hinely, and Meth 2008).

In the following month, the Albert Park area witnessed new signs of antagonism directed towards immigrants. In one attack, local South Africans entered Jamba House on St George's Street and proceeded to evict all the foreigners. Jamba House had been the scene of several police raids. Residents allege that police would simply confiscate whatever they had. Lawyers for Human Rights (LHR), an independent human rights NGO, had taken up a number of cases after residents produced receipts for the goods. Once the residents were out on the streets, police harassed them until they finally disappeared into the night.

Zimbabwean migrant Eugene Madondo, aware of the situation, felt uneasy but thought it was simply the consequence of the excesses of the festive season. As dusk fell, he took refuge in his room at Venture Africa, another nearby block of flats that mainly housed migrants.

'Then came this attack on January 4th, 2009. It was on a Sunday at half past 10 pm. I saw a group of people carrying the weapons like bush-knives and knobkerries. They were carrying hammers too. They were blowing the vuvuzela. They were chanting the slogans in Zulu languages, and some of them were singing. I saw them coming in the direction of the flat where I stay, Venture Africa' (Desai and Madondo 2009).

To get to Venture Africa, the mob had to march past the police station. Closed-circuit television footage caught the marchers on the streets. An employee at the LHR office, close to Venture Africa, repeated that people 'heard the mob, saw the mob'. The owner of Venture Africa was contacted, and he phoned the police to intervene. The police told him the mob was the legitimate work of the Community Policing Forum (CPF). Here is Eugene Madondo's account of what followed:

'[The mob] forced their way up. Six floors.... When it started, I was sleeping, but I was SO scared. I heard the noise of the people screaming. The doors were being broken. They were hitting the doors with the hammers. When I noticed they were even breaking the doors, I thought of running away, so

I opened the door, only to see that my neighbour was being thrown from the sixth floor – DOWN! Before I could lock the door of my room, I was approached by this group of people – a lot of them; there were a LOT! The men were in front, and the women were behind shouting, "Shaya! Shaya! [Beat them! Beat them!]" They asked me where I'm from. I told them I was from Zimbabwe. They shouted, "Shaya kwere-kwere! Shaya! [Beat the foreigners!]" One of them hit me on the head with a knobkerrie. The blood started to flow. I felt weak. I screamed for help. My attacker was joined by other guys. They hit me with blows on my stomach. They even hit my genitals. I felt so weak. I fell down. I thought it was over. I heard one of them talking in a Zulu language. They said, "Let's throw this dog outside the window" in Zulu. Five of them, they lift me up, pushing me through the window. I tried to hold the window frame, calling for help. They broke the window frames, pushing me down from the fifth floor. Lucky enough, I fell on top of these two other guys who were already dead on the floor. When I was down there, and they saw that I was not dead, they carried on throwing stuff and hitting me with empty beer bottles and all those objects. I didn't even notice these guys were dead. I told them, "Come on guys. There's a storeroom nearby that we can crawl to." The blood was coming too much from my head. I could feel the pain in my spinal cord and my head. I lost my conscious[ness].'

The two dead friends that Madondo landed on were Omar Said from Somalia and Victor Zowa from Zimbabwe. After a lengthy stay in the hospital, Madondo made his way back to the scene of death. His room had been ransacked. Many of his fellow tenants had vanished, fearing another attack. Madondo sought help from LHR and the Refugee Social Services. The first thing he was asked to do was go to an identification parade.

'I went there, alone. Most of the witnesses who are not victims are scared. They have fear. They think they will be killed by the perpetrators of the attacks, because most of them are out on bail. On May 19th was the Investigation Parade where I managed to point to some of the perpetrators, including Vusi Khoza, himself. Vusi, what can I say about Vusi? He is so evil. He is lacking a sense of living together. Before the attack he used to come to Venture Africa pointing the fingers [at] the foreigners who live in the flats ... He is not good,

that man. And, he doesn't want to repent, or apologise. He is trying by all means to destroy the evidence by killing the state witnesses.'

'When I pointed him out in the ID parade, he could not see me because of the screen. He was in position fifteen. They said, "Number fifteen, please step forward." When he did that, even though he could not see me, he shook his finger to let me know to feel threatened and have bad feelings. But, I don't hate him. All I don't like is how he treats other people, especially the foreigners. All I need is for him, especially him, Vusi, to be trialled for what he did and he must get a sentence ... The situation at home is causing us to leave our home countries. So, if you take refuge in another country and you find out that the situation in that country is even worse than back home, and local people are out to kill you, then where actually do we go?'

Word was out that Madondo was in Port Shepstone. In fact, after the attack he continued to live in the Albert Park area. It was an act of defiance. Madondo survives in part by sacrificing his family life: His wife, Fungai, and young son, Ashley, have long gone back to Zimbabwe. As the date for the case approached, Madondo went into protective custody. He relied on the protection of the very police who had harassed him during his stay in the city. Khoza, one of the alleged central instigators of the January attack, was also on the prowl in the Albert Park area. On June 26, 2009 the eThekwini Municipality sent out a notice advertising an African Unity Six-a-Side Soccer Tournament, to be held under the banner 'Fighting Xenophobia and promoting African Unity through Soccer'. The letter was signed by Khoza. Throughout 2010, he remained ANC ward councillor and chaired the Metro Police Civilian Oversight Committee, though in the 2011 election he was dropped from the ANC, and in June 2011 his trial began. It was postponed after a day of hearings, as Durban's wheels of justice remain rusty.

In the limited space of the inner city, tensions accumulate, concentrate, and erupt. The CPF has equated crime with the presence of African immigrants. Using this institutional base, they have raided flats where immigrants live and have thrown them out. It has been revealed that slum landlords are charging R700 (around US$100) a month for a small subdivided space, often without water or electricity, in what were once warehouses and dilapidated office blocks. Once immigrants are on the streets, the police move in to harass and force them to keep moving, until finally they leave the area altogether. According to the LHR, the police have taken to raiding flats and simply confiscating immigrants' property, regardless of whether they have receipts of purchase or not.

A block east of Albert Park, refugee support organisations at the Diakonia Centre (a multi-faith site of progressive activism for several decades) have their hands full. Both the LHR and the Refugee Support Services are contracted by the UN High Commission for Refugees. On any one day, the queues are long and resources short. Their work is about papers, legalities and treating each case in an individual way. Some of the organisations were opposed to the concentration of the 47 DRC immigrants at Albert Park. They could not fathom the strategy of the immigrants: How their stubborn presence exposed the lack of support from the city authorities; how their tents staked out a place in the heart of the inner city; how their continuing resistance in the face of their tents being pulled down and harassment by the police, drew attention to ongoing official xenophobia; and how their black plastic bags, which provide them shelter, further exposed the lack of state support. The UNHCR-supported NGOs interpreted the Congolese refugees' actions merely as a game that the immigrants played: Demanding a camp for the sake of facilitating their rendition to a 'first world country'. In their view, the rudimentary shelter was not an act of desperation and fortitude, but one of opportunism and conspiracy.

While support for refugees focuses on technicalities and individual cases, the city uses professionalism to deal with them. KwaZulu-Natal Refugee Council leaders, Baruti Amisi and Pierre Matate (2009), point to the fact that, in its early years, the commemoration of World Refugee Day was organised by the refugee community. However, for the last two World Refugee Days, service providers supported by the government, took over the work of the refugee community. This changed the complexion of the commemoration 'into one for the poor and the refugees (often without food or drink, and usually in a tent); and another one for the agencies in comfortable venues such as City Hall, with expensive food and drinks'. As Amisi and Matate (2009) report, the service providers developed a commemoration 'to celebrate themselves'.

The killings in January 2009 were the acme of high-intensity xenophobia in Albert Park. To some extent, the court case reduced tensions, since it made headlines and put on notice those who are keen to attack foreigners that there could be a price to pay for overt attacks. But there is still a low-intensity threat on the streets, one that emerged again in July 2010 when new xenophobic attacks were reported across the country, after the World Cup. The attacks continue regularly in the course of service delivery protests that turn xenophobic. Yet despite all this violence, Thabo Mbeki repeated his denial of xenophobia in October 2010:

> 'So I am saying that if there was xenophobia, I would expect it to be expressed against people who might stand out as being different from me and also, given our history, these are the people that oppressed us. But you don't have any evidence of racism among our people … Let us get to the root causes [of] this thing and communicate it to the rest of the continent, which I am certain … would confirm the statement that our people are not in the grip of xenophobia' (Ncana 2010).

Similarly, wrote ANC spokesperson Jackson Mthembu (2010) as the World Cup wound down,

> The reported xenophobic attacks by South Africans on foreign nationals, particularly from the African continent, after the conclusion of the 2010 FIFA World Cup in South Africa, is baseless and without any rational [sic].

ANC National Chairperson Baleka Mbete remarked, 'These reports are irrational, have no basis whatsoever' (Mthembu 2010). Meanwhile, in a semi-gentrification strategy, Durban's housing department (notoriously corrupt, as witnessed in the case of the Mpisane tenders, which resulted in enormous profits for the developers of low-quality housing and also an investigation into possible fraud involving R3.5 billion [around US$500 million]) is supporting plans to revamp blocks of flats into sectional title units, that will sell from between R300 000 and R400 000. African immigrants, without access to papers and credit, will not be able to take advantage of these developments. Furthermore, the relocation of the Metro Police Headquarters, in 2010, made Albert Park increasingly inaccessible to immigrants. It is one of the last of the common spaces where African immigrants can meet, lie in the sun, have a view of the harbour, and receive a meal from individual do-gooders or faith-based organisations.

An 'African city' without small traders: A port without fisherfolk?

This process of squeezing out undesirables also included a near-fatal threat to two other vulnerable groups: fisherfolk on the beachfront and informal traders at the Warwick Avenue Market (also known as the Early Morning Market) which, for a century, has been the home of African and Indian fresh produce traders, bovine head cleaners (for the Zulu specialisation of preparing a cow's facial meat) and a host of other informal traders (for a history of the area, see Maharaj 1999). The Warwick Avenue Market has seen battles over space spanning many decades. Official hostility peaked on June 18, 2007 when 500 traders were arrested for minor infractions on a single day. By 2009, city authorities announced plans to replace Warwick Avenue's informal markets with a shopping mall, that was to be constructed by one of the crony business partners – a practice for which Durban has become notorious. Caroline Skinner (2009) puts the struggle in context:

> The market has 673 trading sites, and over 50 market gardeners sell their produce there. These traders supply other informal traders throughout the city, and are thus an important part of the city-wide fresh produce distribution chain. On a busy day, it is estimated that as many as 8 000 street and market traders operate out of Warwick. There are many others working to support this trade, such as barrow operators and suppliers of tables, paraffin and shopping bags. There was no call by the council for expressions of interest when this valuable public land was to be released, nor was there a public tendering process. Also, no existing traders are included in the share ownership of the Black Economic Empowerment consortium driving this development.

The city used various arguments to justify the decision to push out traders, including ridicule by City Manager Mike Sutcliffe of local academics for supporting the market. Why would anyone want to go to such an area, Sutcliffe (2009) asked?

> 'For most of the day the area is dirty, there are far too few ablutions and commuters are very vulnerable, with this being the most dangerous set of intersections in South Africa. There are no banks, post offices and sufficient public services in the area to serve the number of people who deserve access'.

Of course it was under his watch that this state of affairs was allowed to prevail and get progressively worse. While bemoaning the lack of finance to develop this critical part of the city, money does not appear to have been an issue for the development of the massively over-budget Moses Mabhida Stadium, the uShaka Point Waterfront Development and the ICC itself, all requiring tens of millions of rands of municipal operating subsidies. The vast new flyover highway roads, that now take most auto commuters over Warwick Junction (constructed with World Cup infrastructure funding) cost more than R1.4 billion (around US$200 million), but municipal facilities have not improved for the more than 460 000 working-class people who line up at Warwick's taxi stands each day (Comins 2009b).

In 2009, traders and their allies put up incredible resistance to the destruction of the market and forced the city to back down on its plans for the mall later that year. But this could be a pyrrhic victory, since the city refuses to allocate any resources to improve the market or make it more accessible. Still, because it is a place where people can purchase inexpensive, locally-grown fresh vegetables, the Warwick Avenue Market has become one of Durban's most important political-ecological nodes.

A few kilometres due east, on the city's beachfront, fisherfolk are restricted to smaller and smaller areas of the coastline.[3] Since the 1800s, they were able to fish the South Pier harbour mouth (400 metres long), according to filmmaker Sjoerd van Grootheest (2010), but in 2005, invoking the International Ship and Port Security code, the US Consulate intervened to have fishing banned anywhere in the harbour on grounds of port security. This was one of many 'restrictions and onerous regulations forced on South Africa by the United States as part of the "war on terror", [which] could bury Durban harbour and associated businesses under oceans of red tape', according to local reports (Ryan 2007). By November 2009, new municipal angling restrictions were put in place, increasing to the point that, by July 2010, fisherfolk were only allowed to fish from one short, ineffectual pier (commonly referred to as the Snake Park Pier) far from the bigger schools. The fisherfolk have not responded well to this threat to their livelihood, reports van Grootheest (2010):

> Granted, the fishermen have built up a reputation, due to the use of drugs and alcohol by some, at times resulting in conflict and violence on the public space of the beachfront and the piers. The fishermen, hardly a solid or tight community (they come from all over and depend on their catch in differing degrees) have a reputation problem.

According to anthropologist Keith Hart (2011), the hostility partly follows from:

> the Durban Municipality's long-term plans for the seafront (associated with the City Manager, Mike Sutcliffe, an ex-Marxist geographer), which were partly served by World Cup related activity … The City plans to demolish

> the hospital and replace it with a luxury hotel or similar. It has encouraged a lot of investment at this end of the beach including uShaka Marine World, the Point Venice-style housing development, a planned yachting marina and so on. The World Cup saw the extension of the promenade to its limit, and this involved demolition of shacks, informal eateries, etc.

There has been a big police effort in the last five years, making the area a lot safer than it was. The idea is that South Beach will become the natural up-market end of the shore near to the city centre, at the expense of North Beach, where there is now a casino. Durban was ranked in the 'top ten family beach resorts in the world' by *Lonely Planet* last year. Its climate is subtropical and particularly attractive in the southern winter, compared with either Cape Town or Johannesburg. The city faces an eroding tax base. Its port is still the largest in the southern hemisphere, but many high-end businesses are relocating to the northern coast around Umhlanga, and the airport is now 35 km north of the city in what was once Zululand. Tourism is the city's main hope for economic growth, and surfing is intrinsic to that.

Tourism would, in fact, be boosted if fishing was permitted at the port entrance and on the three long piers adjacent to the main hotel strip, say angling advocates. But other battles over space have unfolded because surfers typically jump from those piers to catch waves, occasionally fouling lines or even getting hooked. As Hart (2010) notes,

> The surfers complain that the fishermen leave detritus (animal, vegetable and mineral) where they were. Durban municipality worries about its status as a surfing destination, but even more about the effect on high-end punters of confronting this "mess".

The city went to great lengths in 2010 to redesign the sandy reef next to the piers, pumping sand in to create better waves. However, in the same year, the redevelopment chaos and alleged arrogance of municipal officials resulted in the city's long-standing world surf tournament, the Gunston 500 (since renamed the Mr Price Pro) to move 25 kilometres north to Ballito. Fisherfolk justifiably complain that the mainly white, upper middle-class surfing fraternity has free use of the piers (no user fees), while anglers pay an annual license to fish.

It would have been a simple effort to arrange for the surfers to pay cleaners – perhaps even the homeless kids from the Umthombo shelter, who have taken up surfing to great acclaim – so that hooks and bait would not litter the piers, but instead, atomistic behaviour prevailed. Rather than share the space, persuasive surfing advocates successfully lobbied to have the fisherfolk banned.

Surfers don't have it all their way, however, and too many have emerged from Durban's beautiful beach with runny stomachs and ear infections. On many a hot day, *E. coli* counts soar, especially during periods of sand-dredging or after heavy rains, which wash polluted water and silt into the four major rivers that flow into the sea: the Umgeni (which supplies the bulk of the city's piped water via the Inanda Dam), the Umbilo and Umhlatuzana (which both drain into the port), and the Umlazi, which runs between the two main refineries in south Durban. The Department of Water Affairs' (2010) latest *State of the Rivers* report about KwaZulu-Natal, demonstrated that each has 'poor' water quality at the point where they course through the vast informal settlements on the outskirts of the city, beyond the city's 'sanitation belt' where Durban officials have decided it is not cost-effective to provide decent toilets (Department of Water Affairs 2010).

Even in the densely-populated shack settlements, the municipality provides very few toilets (one per thousand residents is not uncommon) for fear of legitimising the illegal occupations. Hence, rainwater washes faeces from informal settlements into the rivers, with *E. coli* counts by national authorities found to exceed 1 million colony forming units (CFUs) per 100 ml on the Umgeni, near Kennedy Road's shack settlement (the safe limit is 130 CFUs per 100ml, Department of Water and Environmental Affairs 2008). In addition to lack of sanitation at the source, according to the city's leading environmental reporter, Tony Carnie (2008a), the municipality itself 'has been singled out as one of the "most significant" polluters of some rivers because of the failure to repair burst sewer lines and poor management at some waste-water treatment plants' (see also Carnie 2008b and 2008c). The problem of high *E. coli* counts in the seawater became so serious that, from 2008 to 2010, Durban's Blue Flag beach status (signifying internationally-accepted water quality and amenities) was retracted, resulting in a national scandal (Pullan 2010).

Conclusion: Climate justice for Durban?

To be chosen as host city for the COP 17 was a remarkable accomplishment, in light of the ecologically destructive nature of the area immediately adjacent to the ICC venue. The much more beautiful city of Cape Town was rejected, according to a guest post on former City Manager Andrew Boraine's blog, because of

> 'the high levels of security required... The CT International Convention Centre falls way behind the ICC complex in Durban. You can lock it down completely and keep the over-the-top protesters well away from the high-level attendees'.

Boraine, formerly a Johannesburg NGO worker who helped Alexandra Township civic associations defend intense, early 1990s protests against apartheid, later became a facilitator of public-private partnerships. 'Cape Town's proposal', he noted, 'took into account the need to be able to lock down certain areas for government delegations and VIPs' (Boraine 2010). But Durban's was more convincing, and in August 2011, Sutcliffe warned environmental activists (Greenpeace in particular) that if heads of state come to Durban for the COP 17, as they did to Copenhagen, not only will there be a lockdown around the ICC, but there would be no negotiations: '[the] city does not move' as quoted in Khumalo (2011). The test will be the Global Day of Action on December 3, when thousands of climate activists march from Curries Fountain to Warwick, past the Grey Street Mosque and the old Red Square (where heroic anti-apartheid mobilisations occurred), to the US Consulate and City Hall, past the ICC, and finally to the beach for a protest rally, concert and civil disobedience.

Just over a decade earlier, at the end of August 2001, a march of 15 000 to the ICC, led by Fatima Meer and Dennis Brutus against the UN World Conference Against Racism WCAR, came close to barging in on the lethargic delegates (Desai 2002). Conference critics argued that no serious UN discussion was underway about reparations needed for slavery, colonialism and apartheid, nor was there adequate UN action against Israeli racial oppression and occupation of Palestine. In addition, a great deal of the 2001 activism was directed at the way South African class apartheid had emerged since 1994, with resulting amplification of racial differences. Ten years on, Durban's contradictions are more extreme than ever.

Ironically, central Durban's urban economic oppression was amplified by an energetic bureaucrat, Sutcliffe, known for once being a 'Marxist geographer' (Tolsi 2008). Under his rule, the road that goes due west from Albert Park through to the bourgeois suburb of Glenwood, was changed from Moore to Che Guevara, but residents regularly defaced the new sign (as would, perhaps, Che himself). Again, ironically, in 2008 the American Association of Geographers awarded Sutcliffe the

Gilbert F. White Distinguished Public Service Honors, and the James R. Anderson Medal of Honor in Applied Geography. Sutcliffe's sponsor for the award, Kevin Cox (a faraway Marxist geographer who supervised Sutcliffe's 1984 Ohio State University Ph.D. thesis) described these awards as

> 'among the most prestigious recognitions in geography ... his training as a geographer has been at the centre of much of his accomplishment. Over a lengthy career as political activist and trusted member of the ANC government, Mike has proven himself to be an applied geographer par excellence and with a strong pro-people bent' (Cox 2008).[4]

The vulnerable people (including children, currently under attack from Durban's neoliberal-nationalist municipal government) are also those urban residents (sheltered in shacks or slum hovels, seeking to make a living at society's margins) who are likely to suffer most under climate change. Defensive measures, such as the Warwick Avenue Market struggle, are only a holding pattern because the origins of the commodity chain (through which fresh fruit and vegetables enter Durban) will be threatened in coming decades by worsening droughts and floods, as will subsistence agriculture in rural KwaZulu-Natal or Zimbabwe's Southern Matabeleland (thus prompting a new migration wave of climate refugees), including the catches of desperate fisherfolk when the ocean warms. Moreover, as the 2008 xenophobia incidents and 2009 campaign against the Warwick Avenue Market's displacement showed, enormous efforts are required for local market activists and immigrants to unite with other constituencies under threat, in no small part, because of decades-old racial and ethnic tensions.

The idea of a broad front of Durban's oppressed appears as a far-off aspiration. And yet, those who seek such an alliance (the organisers in the South Durban Community Environmental Alliance [SDCEA] and other groups, such as the 'Durban Social Forum' and Climate Justice Now! KwaZulu-Natal) are continually reminded of city leaders who surprised powerful government and business forces during the 20th century, and often made enormous progress: Mahatma Gandhi's *Satyagraha* strategies and tactics in Durban's Phoenix township, prior to departing for India in 1914; the ANC's founding president John Langalibalele Dube and its first Nobel Prize-winning president, Albert Luthuli; the martyred communist leader, Johannes Nkosi; Monty Naicker, who in the late 1940s allied Indians with the ANC; the founder of the Black Consciousness Movement, Steven Biko; United Democratic

Front leaders Victoria and Griffiths Mxenge; the 1973 dockworkers; the Chatsworth community activists who revived class politics in the late 1990s; those who built the Diakonia Council of Churches liberatory legacy; and intellectuals like the assassinated political theorist, Rick Turner and more recently, sociologist Meer and poet Brutus (now dead, but remembered with enormous regard by activists). Occasionally, as at the WCAR in 2001, there is hope that these traditions of Durban activism find like-minded opportunities for solidarity with visitors.

If they were with us, what would the most advanced of these activists and strategists be doing in central Durban in late 2011? What would their demands be? How would they approach the COP 17? They would probably consider, with utmost seriousness, the world Climate Justice movement's Cochabamba People's Agreement of April 2010, adopted at a conference of 35 000 in Bolivia. The core demands in that document are as follows:

- reduce greenhouse gas emissions by half by 2017 to stabilise temperature increases to 1 degrees Celsius and 300 parts per million of CO_2
- acknowledge the climate debt owed by developed countries (6 percent of GDP per year)
- respect indigenous people's human rights and the rights of Mother Earth
- establish an International Court of Climate Justice
- reject carbon markets and commodification of nature and forests through REDD
- change consumption patterns of developed countries
- end intellectual property rights for technologies useful for mitigating climate change.

These demands cut against the grain of the UNFCCC at least four times: Firstly, far greater emission cuts are required than the present balance of forces in negotiations will permit; secondly, the UNFCCC's failure to take climate crimes and the climate debt owed to victims of climate change seriously; thirdly, the UNFCCC's commodification of everything, from intellectual property to forests (and charismatic species within them); and lastly, the UNFCCC's failure to consider decommissioning the dangerous carbon markets. While the closest that the UNFCCC will come (in Durban) to a deal on financing is clarification on the disappointing Green Climate Fund, there is every likelihood that the worst possible outcomes of compensation monies (originally pegged at US$100 billion per year) will be realised, insofar as carbon markets will fill half the fund (with all that that entails), and decisions will be made to press ahead with financing 'false solutions', instead of committing to the deep emissions cuts and economic transformation that is required to solve the crisis.

Durban's hosting is why the most powerful memories from the COP 17 could be very local; the city itself confirms that a prefigurative politics of injustice is already playing out in the oppression of so many people in the vicinity of the ICC, leaving us as pained as Herbert Dhlomo and as irritated as Dennis Brutus about the uncertain liberals who continue to legitimise the UNFCCC as its delegates polish the chains of global climate apartheid. And it is why those international climate activists (who correctly worry about losing the last chance to keep the binding emissions-cut requirements of the Kyoto Protocol alive), will likewise be furious in December when their fears are realised. The question is whether these kinds of people will find common cause, as have so many others in Durban.

References

Academy of Science of South Africa. 2011. *Towards a low carbon city: Focus on Durban.* Pretoria, http://www.assaf.org.za/2011/08/durban-on-a-pathway-towards-a-low-carbon-city/#more-4981.

American Association of Geographers. n.d. Anderson Medal of Honor, http://www.aag.org/cs/awards/anderson.

Amisi, B., P. Bond, N. Cele, and T. Ngwane. 2011. Xenophobia and civil society: Durban's structured social divisions. *Politikon* 38 (1): 59–83.

Amisi, B. and P. Matate. 2009. Refugees call for unity and an end to xenophobia. *The Mercury*. June 24.

Bass, O. 2009. Aiming for Africa: Durban 2010 and notions of African urban identity. In *Development and dreams: The urban legacy of the 2010 football World Cup*, U. Pillay, R. Tomlinson, and O. Bass, eds., 246–265. Cape Town: HSRC Press.

Bond, P. 2002. *Unsustainable South Africa.* London: Merlin Press.

———. 2010. Maintaining momentum after Copenhagen's collapse: "Seal the deal" or "Seattle the deal?" *Capitalism Nature Socialism* 21 (1): 14–27.

———. 2011. From Copenhagen to Cancún to Durban: Deckchair shifting on the climate Titanic. *Capitalism Nature Socialism* 22 (3): 3–26.

Bond, P., R. Hinely, and O. Meth. 2008. Great White Shark mauls Albert Park refugees. *The Mercury.* November 12.

Boraine, A. 2010. Cape Town loses bid to host UN climate talks. Andrew Boraine blog post. November 12, http://www.andrewboraine.com/2010/11/cape-town-loses-bid-to-host-un-climate-change-conference-in-2011/.

Carnie, T. 2008a. Water courses awash with sewage: Safety threat to city rivers. *The Mercury*. January 31.

———. 2008b. High levels of *E. coli* found in Durban rivers. *The Mercury*. January 31.
———. 2008c. Durban beaches under threat: Red flag for blue flag beaches. *The Mercury.* March 14.

Comins, L. 2009a. Street kids "won't disappear" for 2010. *The Mercury.* April 2.

Comins, L. 2009b. Traders feel threatened by development. *The Mercury,* 20 May.

Cox, K. 2008. Kudos! *Geospectrum 2008*, http://www.geography.osu.edu/docs/geospectrum/geospectrum.pdf.

Crush, J., D. McDonald, V. Williams, K. Everett, D. Dorey, and T. Taylor. 2008. *The perfect storm: The realities of xenophobia in contemporary South Africa*. Migration Policy Series 50. The Southern African Migration Project. Cape Town: Idasa and Kingston, ON: Southern African Research Centre, Queen's University, Canada.

Dardagan, C. 2011. Stadium shortfall is minimal, says Sutcliffe. *The Mercury*. August 1.

De Boer, H. 2008. New policy proposed for Durban beaches. *Daily News*. March 14.

Desai, A. 2009. Interview with Eugene Madondo in Durban. June.

Desai, A. and S. Walsh. 2010. Knowledge and power in South Africa: Xenophobia and survival in the post-apartheid state. In *Global perspectives on social movements and knowledge production: Learning from the ground up*, A. Choudry and D. Kapoor, eds. New York: Palgrave Macmillan.

Deibert, M. 2008. A glittering demon: Mining, poverty, and politics in the Democratic Republic of Congo. *Corpwatch*. June 26, http://www.corpwatch.org/article.php?id=15118.

Department of Water Affairs 2010. *State of the Rivers: Umngeni,* Pretoria, http://www.dwa.gov.za/iwqs/rhp/state_of_rivers/state_of_umngeni_02/umngeni.html.

Godsell, B. 2005. Doing business in a conflict zone. *Business Day*. June 3.

Hart, K. 2010. Comment. Urban Anthropology blog. http://openanthcoop.ning.com/group/urbananthropology/forum/topics/the-exclusion-of-fishermen?commentId=3404290%3AComment%3A83322&xg_source=activity&groupId=3404290%3AGroup%3A1436.

Hewitt, T. 2010. Durban and beyond. *Umthombo News*. January, http://www.umthombo.org/site.php?level=8&id=80.

Human Rights Commission (HRC). 1999. *Report on the arrest and detention of persons in terms of the Aliens Control Act*. Johannesburg: Human Rights Commission.

Johwa, W. 2008. Warning on xenophobia ignored. *Business Day.* June 8.

Khumalo, S. 2011. Durban ready for mega-event. *Daily News*. August 17.

Laganparsad, M. 2010. Metro police deny charges of *street children* abuse. *Sunday Times*. February 14.

MacDonald, D. 2004. The political economy of identity politics. *The South Atlantic Quarterly* 103 (4): 629–657.

Magwaza, N. 2008. Metro police told to "move refugees." *The Mercury*. November 3.

Maharaj, B. 1999. The integrated community apartheid could not destroy: The Warwick Avenue Triangle in Durban. *Journal of Southern African Studies* 25 (2): 249–266.

Mail & Guardian. 2010. Metro police round-ups of street children in Durban. January 21, http://www.mg.co.za/article/2010-01-22-rounded-up-and-shipped-out.

Mattes, R., D.M. Taylor, D. McDonald, A. Poore, and W. Richmond. 1999. Still waiting for the barbarians: S.A. attitudes to immigrants and immigration. *Southern Africa Migration Project* 14. Cape Town.

Mdletshe, C. 2008. Go back home, immigrants told. *The Sowetan*. November 6.

Mthembu, J. 2010. No outbreak of xenophobia looming. African National Congress press statement. Johannesburg, July 2, http://www.politicsweb.co.za/politicsweb/view/politicsweb/en/page71619?oid=184535&sn=Marketingweb+detail.

Murray, M. 2003. Alien strangers in our midst: The dreaded foreign invasion and "fortress South Africa." *Canadian Journal of African Studies* 37 (2/3): 440–466.

Ncana, N. 2010. Mbeki calls for probe into attacks. *Sunday Times*. October 17.

Nyamnjoh, F. 2006. *Insiders and outsiders: Citizenship and xenophobia in contemporary Southern Africa*. Dakar: Codesria Books.

Packree, S. and H. de Boer. 2007. Where are Durban's street children? *Daily News*. November 22.

Potts, A. 1988. Picturing the modern metropolis: Images of London in the nineteenth century. *History Workshop Journal* 26: 28-56.

Pullan, G. 2008. Poor quality beaches will be their legacy. *North Coast Courier*. September 10, http://www.northcoastcourier.co.za/2010/09/08/10-september-4/.

Reuters. 2008. "Congo war-driven crisis kills 45,000 a month-study." 22 January, http://www.reuters.com/article/2008/01/22/us-congo-democratic-death-idUSL2280201220080122.

Robinson, I. 2010. Shame on the beautiful game, http://redcardcampaign.wordpress.com/2010/06/11/shame-to-the-beautiful-game-video/.

Ryan, M. 2007. Red tape to choke Durban port. *Sunday Tribune*. October 21.

Skinner, C. 2009. Warwick Mall crisis for traders and Durban's political vision. *The Mercury*. May 27.

Sole, S., 2010. Durban's Moses Mabhida Stadium: Arch of hope or yoke of debt? In *Player and referee: Conflicting interests and the 2010 FIFA World Cup*™, C. Herzenberg, ed. Pretoria: Institute for Security Studies.

Sutcliffe, M. 2005. Street children. *City Manager's Newsletter*. Durban. May 13.
———. 2009. Response to academics around Warwick developments. *City Manager's Newsletter*. Durban. June 4.

The Mercury. 2010. Book exposes deals behind Durban stadium. April 30, http://www.iolproperty.co.za/roller/news/entry/book_exposes_deals_behind_durban.

Tolsi, N. 2008. Party tricks and power play. *Mail & Guardian*. August 7.

Van Grootheest, S. 2010. The exclusion of fishermen from the sea. Post to urban anthropology discussions. November 12, http://openanthcoop.ning.com/group/urbananthropology/forum/topics/the-exclusion-of-fishermen?commentId=3404290%3AComment%3A83322&xg_source=activity&groupId=3404290%3AGroup%3A1436.

Wikipedia. 2011. eThekwini, http://en.wikipedia.org/wiki/EThekwini.

Endnotes

1 Simmer previously worked for Remnant Alton, the firm that failed miserably in its attempt to privatize Durban's municipal bus service. He had also served as director of the ill-fated Dolphin Whispers development, luxury apartments at the Durban Point Waterfront (*The Mercury* 2010).

2 This level of intervention was not uncommon. At Durban's main beach in July 2010, police officers who arrested one author for distributing an anti-xenophobia leaflet before the World Cup's Ghana-Uruguay match (termed 'ambush marketing') confirmed in a taped conversation that the City Manager had explicitly ordered: 'No distribution of pamphlets, especially which mention xenophobia.' The reasoning, according to a police superintendent, was that '[y]ou are reminding [people] of xenophobia' (Bond 2010).

3 For more on the fisherfolk's struggle, see http://vimeo.com/10840958 and http://researchspace.ukzn.ac.za/xmlui/handle/10413/276.

4 According to the Association of American Geographers (n.d.), the Anderson Medal of Honor reflects 'the most distinguished service to the profession of geography'. The idea of 'Public Service', is of inestimable value to the community. Leadership in the wide spectrum of community affairs offers exceptional opportunities to the geographer to earn distinction. This effort should have been sustained over a period of time, have gained more than usual recognition by co-workers, public officials and fellow citizens, and have clearly influenced the progress of the community.' (AAG n.d.)

'More than usual' is correct. Wikipedia's entry on Mike Sutcliffe begins as follows:

> Michael Sutcliffe is municipal manager of the eThekwini Metropolitan Municipality (population 4 million), which includes the city of Durban, South Africa. As well as being the municipal manager for the region, he is also currently the most unliked person in the greater eThekwini area. Deployed to the position in 2002 by the ruling African National Congress, he oversaw the municipality's successful preparations for the 2010 FIFA World Cup, and was involved in controversy regarding street renamings, the loss of the city's Blue Flag beach status, illegally banning protests, banning posters, serious human rights abuses in the city's housing program, the failed privatization of the city's bus system, allegations of spin-doctoring, the failed uShaka Marine World, threats to withdraw advertising from newspapers employing journalists critical of the municipality, lack of action against environmental destruction, favoritism toward ANC-aligned individuals and businesses, unlawful and at times violent violations of the basic rights of street traders and shack dwellers and corruption. According to the Sunday Times he is about to be arrested on corruption charges. (http://en.wikipedia.org/wiki/Mike_Sutcliffe, accessed August 10, 2011).

On the latter charge, according to a mid-2011 report, "The [regional] African National Congress called on [provincial local government minister Nomusa] Dube to order a forensic investigation after the auditor-general found that the city had irregularly spent US$80 million, and the audit implicated Sutcliffe and three officials in irregular housing contracts worth US$500 million over 10 years" (Nair, 2011).

Chapter 2

Between Apartheid and Neoliberalism in Durban's Indian Quarter

Ashwin Desai and Goolam Vahed

A drive from Durban's beaches towards the once-bustling Indian Quarter will lead to a confluence of three streets: Monty Naicker, Dr A B Xuma, and Dr Yusuf Dadoo. It is appropriate that they meet at a point that was once called the Red Square (and is today the site of the Nicol Square multi-storey parking garage). This always-congested traffic intersection was the venue for many an anti-apartheid rally during the tempestuous decade of the 1950s, a period that first brought the likes of Nelson Mandela, Albert Luthuli, Dr Yusuf Dadoo and Helen Joseph fully into the pages of history. Most significantly, Red Square was the gathering point for the historic 1952 Defiance Campaign, the first organised joint Indo-African opposition to apartheid.

The threesome of Dr Alfred Bitini Xuma, leader of the African National Congress (ANC); Dr Yusuf Dadoo, president of the Transvaal Indian Congress (TIC); and Dr Monty Naicker, president of the Natal Indian Congress (NIC), were the pioneers of the very first attempt in the city to unite African and Indian struggles against racial oppression. The declaration of cooperation between their three organisations (the NIC, the TIC, and the ANC), signed in 1947, is recorded in the history books as the 'Doctors' Pact'.

Indians began arriving in Natal in 1860, and by the 1890s, the Indian Quarter had already taken shape as the authorities used the 'sanitation syndrome' to segregate Indians from the white boroughs. Indians were forced to settle in three zones around Durban: a business- and commercial area bounded by Pine Street (now Monty Naicker Road), Albert Street, the railway line and the cemetery on Dr Pixley Kaseme Street (previously known as West Street); an area around Garnet Road and bordering Umgeni Road; and the area between Johannes Nkosi Street (previously known as Alice Street) and the Greyville Racecourse (Warwick Triangle).

Growing up in the Indian Quarter

We grew up in the Indian quarter during the 1960s and 1970s, Vahed on Pine Street (today, Monty Naicker Road) and Desai on Prince Edward Street (today, Dr Goonam Street). Our histories are folded into the labyrinth of this area. Writing these words is to capture our boyhoods on paper. It was a time of street football (illegal, of course); the electric atmosphere on Saturday nights as thousands thronged Victoria Street – dressed to the nines – to watch English and Hindu movies at the Shah Jehan, Avalon, and Naaz cinemas; the adolescent male, street corner society with its petty chauvinisms and repressed sexualities; the gangsters whom we both feared and respected.

We were witnesses, too, to the emptying-out of the quarter: the movie houses closing; neighbours across generations disappearing into the designated Group Areas on the outskirts of the city; the Quarter locking down every night as shop owners pulled down their steel shutters. It was a sign of the times that even the mosques had to shut shop. Those who survived the Group Areas were relocated when the Western Freeway was built in the 1970s. It comes into the city at the very point where Red Square was located. In the name of *development*, even the dead were not spared. Part of the cemetery was lost and several hundred graves were dug up, much to the chagrin of locals.

Diagonally across the road, to the right of the old Red Square, are the Ajmeri and Madressa arcades. They are, in reality, tiny lanes filled with a multitude of shops whose goods spew out onto the narrow pathway. One wonders if this is not a canny way to ensure that informal traders do not trespass! Old crafts and memories harking back to the past still linger. "Record King" plies a trade in vinyl with copies of The Temptations and Earth, Wind and Fire with tattered original covers snatched up at basement prices by aficionados. There is one watch repairer left in the arcades, squeezed into a shop window, seemingly oblivious to the times we live in. It was here that housewives once bargained for quality 'India pots', *aghar batti* (incense), plastic utensils and intricately-designed gold jewellery. It was here, too, that the most sought-after tailors (*darjees*) got the measure of their clients, and many a father bought his son the popular handcrafted carrom boards on which to play finger billiards – a wonderful pastime before the age of the internet and cellphones.

It was in these arcades that young Indian leftists of the Communist Party (CP) studied the political economy of the world in the 1930s and 1940s, and swore allegiance to both Stalin and Gandhi, as they plotted the overthrow of the conservative grandees of Indian politics, who preferred appeasing the minority white regime instead of confronting it. It was also in these arcades that many a young couple first held hands in the anonymous throng, and felt the anticipation of more beyond the gaze of orthodox parents and aunties and uncles.

The 1950s saw an explosion in the city's nightlife. For the first time, Indian women were seen dancing, smoking and having a drink at the Himalaya Hotel on Beatrice Street (today, Charlotte Maxeke Street) and Goodwill Lounge on Victoria Street (today, Bertha Mkhize Street), where overseas and local jazz musicians performed into the early hours of the morning. The Delhi Restaurant at the corner of Pine (Monty Naicker) and Dr Yusuf Dadoo Streets competed with Point Road (today, Mahatma Ghandi Road) as a place of 'sin', reeking of alcohol and cheap perfume.

Walk from Dr Yusuf Dadoo Street, through the Ajmeri or Madressa arcades, and you will come to the appropriately named Cathedral Road. Turn right and you will be faced with the mighty Catholic church, St Emmanuel's Cathedral. But in the cheek-by-jowl multi-cultural give-and-take that the Indian Quarter has become, you will first have to pass the iconic Manjra's Café, where all classes lined up (and still do), for its signature dhal and rice. It is only for those with a strong constitution, for the dhal can cause tummy rumbles.

Sacred spaces

It was at St Emmanuel's Cathedral that a youthful Denis Hurley held sway, firing sermons for almost six decades against segregation, while liberation Catholicism rocked to the 5:30 Saturday afternoon African mass. Next door, and occupying the whole block on Dr Yusuf Dadoo Street, is the Grey Street mosque (Juma Musjid), the largest mosque in the southern hemisphere. Muslim traders Aboobaker Amod Jhaveri and Hajee Mahomed Hajee Dada built the mosque in 1881, with the resplendent minarets added in 1904. The mosque represents much of the past and the new. The past is reflected in the composition of the trustees who control the mosque: following a century-old tradition, they are still selected on the basis of ethnicity related to region of origin in India. The new is witnessed in the composition of the congregation, which now has a majority of migrants from South Asia and Africa, reflecting important trends in the post-apartheid South African landscape.

Wealthier Muslims, having evacuated the central business district, have built mosques closer to their mansions in the suburbs. It is a sign of the times that the Himalaya Hotel, where Chief Albert Luthuli was given a thunderous farewell lunch en route to Oslo, to collect his Nobel Peace Prize in 1961, has now been converted to a mosque frequented by the few South African-born Indian Muslims still left in the city. Above the mosque, the hotel has been revamped into apartment blocks with large 'For Sale' boards plastered on the walls. There are few takers as nostalgia fails to ignite a return to the city.

The old congregation at St Emmanuel's Cathedral has all but evaporated. The whites of the inner city have fled into the gated suburbs of the North Coast or Australia. Their domestic workers, who filled the pews of the African mass on Saturdays, followed the labour market out of the city. The Indian Catholics, many of whose forebears arrived as indentured labourers from locations like Pondicherry (where there was a strong French influence), and who now lie buried behind the church, were bundled off to Chatsworth and Phoenix in the grand apartheid sweep of the Group Areas Act. Many of their children have sought redemption in Pentecostalism. The church, though, remains socially conscious. The Cathedral was a hive of activity during the anti-apartheid struggles of the 1980s, and is now a shelter for African refugees. Despite its age, St Emmanuel's Cathedral, which was built in 1904, remains an impressive building architecturally. The cross depicting Christ's crucifixion was carved out of Italian Carrara marble, gifted by the wife of Napoleon III of France, Empress Eugenie, in memory of her son, the Prince Imperial, who was killed in the Zulu War of 1879.

Flavours of home

Gujarati-speaking Hindus made up a large segment of the Indian trading class, and they had a strong presence in the Indian Quarter, introducing vegetarian restaurants (Kapitan's and Patel's being the most famous) and buildings, like the Surat Hindu Association, constructed in 1907 on Prince Edward Street (today, Dr Goonam Street). How wonderful that Dr Goonam is recognised in this fashion: This whisky-swilling, cigarette-smoking 'Coolie Doctor' (the title of her magnificent autobiography) was the first female Indian doctor in Natal, and fought apartheid bravely, but refused to bow to the liberation movement's conventions. It is appropriate that she gets to replace British Royalty. When the Royal Family visited South Africa in 1947, she led the call for a boycott. The Indian community, however, turned out in the thousands to hail *their* Queen. Dr Goonam may have lost that contest but won the war, as history remembers her courage.

No visit to the Indian Quarter is complete without a 'quarter beans' or 'quarter mutton' at one of the many Indian cafés and restaurants in the area, Patel's Vegetarian Lounge, Victory Lounge, and Bhagat's have superseded Kapitan's, the pre-eminent Indian eating house from the 1920s to the 1980s. Indian Delights, such as *bajjia, sev and nuts* (a spicy mixture of green lentils and gram flour noodles), and colourful sweetmeats dot the windows. But the 'bunny chow' holds sway. It is a home-grown fast food and traditionally consists of a quarter loaf of bread, from which the top portion is cut off, the hollowed part is filled with a curried stew of one's choice, and then covered with the portion that was cut off the top. In the early days, the curry of choice was beans, and broad beans were the all-time favourite. Eating a bunny chow is typically a communal activity, as groups of two or more persons usually *gammal* (eat) together. The name probably emanated from the fact that Kapitan's and Patel's on Grey Street (now Dr Yusuf Dadoo Street) were owned by *Banias* (an Indian trading caste), who created this dish as a cost-effective and convenient take-away. *Banias* were also referred to as 'bunny ou's'. ('Ou' is the Afrikaans equivalent of 'guy'.) Today, the bunny chow is also on the menu of many five-star establishments. It is one of those side-splitting, curry-spilling moments to see *lahnee* (rich) Indians tucking into a bunny chow (mutton curry, more than likely) in one of those places, with a fork and knife.

Today, new migrants from India and Pakistan are threatening to dominate the old Indian food outlets. Chicken *tikka* and foods long-forgotten from Gujarat have re-appeared. Kapitan's has disappeared, and in its place is Little Gujarat. Locals have responded with 'real Durban curry'. But it is a losing battle, not only because of the taste, but also the price. A beautifully made, delicious *kadhi khichdi* (yellow rice with a spicy, yellowish yoghurt sauce) at Little Gujarat costs R10. It is the *de facto* food of comfort for many. A *real* Durban vegetable curry costs between R30 and R50.

Indian Quarter: Then and now

The Indian Quarter was once an area in which shop windows were graced with mannequins, resplendent in beehive hairdos and the latest saris. And dotted around them were 'material shops'. A few of the shops remain, but their colour has long faded. Today, the shoppers are mainly working-class Africans shopping for bargains in the new, mainly Chinese and Pakistani-owned outlets. At the intersection of Ingcuce Road (the old Albert Street) and Dr Goonam Street (the old Prince Edward Street), in the 'Pakistani Quarter', are many Pakistani hairdressers. For R20, a fifth of the cost of a haircut in the malls, one can get a haircut, shave and head massage. A treat not to be missed! The *paan* (betel leaf) is thrown in as a bonus. Bargaining remains central to the shopping experience. There are several restaurants serving authentic Pakistani delicacies, with *tandoori* chicken *braaied* (barbecued) on street corners. The aroma of exotic spices, ginger and garlic is tormentingly mouth watering.

Opposite the mosque is the Gandhi Library, standing as a stark reminder of Mahatma's influence on the city, and the only reminder in the Indian Quarter. Ironically, Point Road, the spine of Durban's red light district, has been renamed Mahatma Gandhi Street. However, it might not be so ironic, given Gandhi's tortured experiments with his own sexuality.

The area carries few memories of the historically important African presence in the area. At one end of Dr Yusuf Dadoo Street is a hostel for African women. This is a throwback to the era of apartheid, when African women were allowed into the city for the express purpose of administering to the needs of white children. It was a single-sex hostel *sans* any family, for the women could not be distracted by the demands of their own families.

Directly next to the Indian Market in Victoria Street (today, Bertha Mkhize Street) was a beer hall for Africans. A century ago, small numbers of Africans lived in backyards and other forms of informal accommodation, while several African traders owned stores in the area. There was an African Market next to the Indian Market where beadwork, Zulu shields (handcrafted from cowhide) and meats were sold, and where herbalists and hawkers plied their trade. Several 'native eating houses' also dotted the area. Some Africans, mainly women, sold beer, until the city council acquired a monopoly on this lucrative trade. The Industrial and Commercial Workers' Union, the largest African union in the 1920s, operated for a while under the leadership of AWG Champion, in premises rented from ME Paruk on Leopold Street (today, David Webster Street).

On Charlotte Maxeke Street, there is the YMCA where the CP held its meetings in the 1930s, and where Steve Biko first preached of black consciousness in the 1960s. Originally started on Queen Street (today, Dennis Hurley Street) as the Durban Bantu Social Centre for the aspiring African middle class, music concerts, dances, soccer games and boxing matches were held here to distract them from the pressing political issues of the day. There is little of this past African influence left, though. Removed from the Indian Quarter by the 1960s, Africans have returned to dominate the area as shoppers and street vendors.

Standing oblivious to the changes and dispossessions, wrought by the developmental bulldozer, is the shrine of Badsha Peer in the Brook Street Cemetery. Sixteen acres in size, the cemetery was opened in 1864 and apportioned to (white) Episcopalians, Wesleyans, Roman Catholics and 'other' Christians, while a mere two acres were set aside for Indians and Africans. This sacred site had to concede ground to 'progress' when part of the cemetery was dug up in the late 1970s to make way for the Western Freeway. The shrine is the home of a Muslim holy man, Sheik Ahmed, who came on the *Truro*, the first ship to transport indentured Indians to Natal in 1860. The saint's *karamat* (miracles) are legendary. They relate to healing the sick, assisting the weak, foreseeing accidents and protecting others.

Warwick Junction includes many important landmarks. Sastri College was the first Indian high school in South Africa. Next to the college is the Curries Fountain Sports Ground. It enjoys an iconic status because of the numerous cultural, sporting, and political events held, since its opening in the 1910s. Today, Currie's Fountain stands as forgotten as the political and sporting ideals of yesteryear.

A short walking distance from the Gandhi Library is the Victoria Street Market, one of two Indian markets built almost a century ago. Its history is really a story of Indian market gardeners and petty traders, who, despite the most callous attempts by the racist city council to erode their access to customers in the heart of the city, refused to cower. On a Saturday, this major tourist attraction drew as many as 50 000 visitors to its more than 200 stalls, which sold everything from curios, trinkets, and antiques to fish, rice, groceries, cooked food (such as *puri patha*) and sweetmeats (such as *jalebi*). From the very beginning, the authorities monitored the market vigilantly and complained of overstocking and unhygienic conditions. 'Safety' then became the issue. Despite pressure from the city councillors, the market endured. When it burned down in 1973, it was rumoured (with good reason) that the white city fathers had a hand in the fire.

Following pressure from the community (largely to boost the image of government-appointed organisations, like the South African Indian Council), Victoria Street Market was resurrected in the 1980s, into a modern oddity with yellow walls and purple minarets. Inside the market, the attempt to catch the exotic feel of the East falls as flat as medieval Catholicism's take on the shape of the earth. Somehow, one gets the sense that the resurrection is a fake. But still, the old favourites are here. Coloured spices shaped in cones tumble out of stainless steel pots. Bowing to the changing market, the curry powder shops do a roaring trade in Chinese balms, to a mainly African market. We know how the ANC government-aligned Congress of South African Trade Unions feels about Chinese T-shirts: they actively oppose the presence of these items in South Africa, because cheap imports have decimated the local clothing and textile industry. But they have said nothing about *this* best-seller. Across from the major spice shop is a table with a plethora of beads, including little ones shaped in the sign to display sensitivity to HIV/AIDS. Hanging above them are sarongs with the face of South Africa's president, Jacob Zuma.

There are also a few curio shops, where one gets a sense that there is always something more going on behind the curtain. In this age of canned hunting, we wonder whether anyone still buys curios of lion heads. We ask who the biggest customers are: 'Germans', is the instant reply. Some Europeans are clearly more sensitive to the ongoing destruction of Africa's wildlife heritage, than others. Inevitably, the ubiquitous Pakistani cellphone shops across the city have a home in the Victoria Street Market. Business in *World Call* phone cards is booming. There are tens of thousands of African and South Asian immigrants in the city. Largely shunned by locals, they seek connections elsewhere.

Despite being hemmed-in on all sides by the latest world-class tourism developments, and neglected and denied funds to upgrade the area, the Indian Quarter, that was honed over a century, simply refuses to die. The mega projects surrounding the Quarter only serve to increase its allure: a Pakistani haircut with a head massage thrown in; vegetable curries sold by the most recent immigrants from Gujarat; a beautifully tailored suit; old gold rolled into new designs. This is a place that apartheid, not from want of trying, could not kill off. It is pushing back against the new development machine. A place where the very last watchmakers are sitting with goggled eyes, scrutinising the last springs deep inside the watches that our fathers wore. This is Durban's Indian Quarter today.

Chapter 3

Ways of (Sight) Seeing Central Durban

Ashwin Desai

To have a sense of the geography of the centre of the city of Durban, the most revealing vantage point is the ridge that runs across from Berea to Morningside. To take in its beauty, the best time is in the late evening when the sunlight is quickly replaced by the city lights. To make sense of the city is to appreciate that colonialism, segregation and apartheid have all left their imprints.

At the heart of this enterprise was an accumulation of white privilege and power – through dispossession. Rosa Luxemburg provides an incredibly powerful analysis of this process:

> The European conquerors are the first who are not merely after subjugation and economic exploitation, but the means of production itself, by ripping the land from underneath the feet of the native population… European capitalism deprives the primitive social order of its foundation. What emerges is something worse than all oppression and exploitation, total anarchy and a specifically European phenomenon, the uncertainty of social existence… Before the advance of capitalism, the primitive social order, which outlasted all primitive historical phases, capitulates. Its last remnants are eradicated from the earth and its elements – labour power and means of production – are absorbed by capitalism (Hudis and Anderson, 2004:110).

Heroic battles were fought by Zulu warriors against colonial policy that was designed to loot the land and turn the Zulu into wage slaves. The most stunning of these battles took place in 1879 at Isandlwana. While the Zulu army was victorious at Isandlwana, the British response ultimately led to the breaking of the Zulu kingdom's power. In many respects, the sacking of the kingdom was to shape the contours of the province and give enormous powers to those who ran the city of Durban. The city became a laboratory of racial exclusion and provided the blueprint of apartheid geography, that sought to drive 'non-whites' to the outskirts of Durban's boundaries.

In 2011, almost two decades after apartheid collapsed, Durban's skyline has changed dramatically. Looking down from the ridge, the city is dominated by a huge arch of lights spanning the brand new Moses Mabhida Stadium.

A tale of two stadiums

Moses Mabhida Stadium is the new rulers' spectacular assertion of the city's power. Built for the soccer World Cup 2010, it is the city's iconic landscape feature that symbolises Durban's commitment to becoming a world-class destination for mega sporting events. Right opposite Mabhida Stadium is another stadium, home to the Sharks provincial rugby outfit. The Sharks Stadium (Kings Park), during the days of apartheid, allowed a few 'non-whites' to squeeze into a separate stand with a single toilet. Very few took up the offer, for this is where one could run afoul of white louts, looking to have some fun at the expense of the 'non-whites'. As the passage below reveals, rugby was central to Afrikaner nationalism, being considered

> a means of national expression of the Volk [Afrikaner nation] … Many seemed to see international rugby as the means of impressing upon the world the existence and toughness of the Afrikaner people, and this approach was undoubtedly the result of seeing themselves as a beleaguered group, special to their circumstances but also special in their mission (Donald Woods quoted in Grundlingh 1996:188).

At Kings Park, Afrikaner Nationalism and the children of the British colonial order came together. 'The rugby ground became a symbol of white power, identity and collective consciousness …'. (Nauright 1997:86).

'Non-whites' did play the game. But it was a different world to what we witnessed at Kings Park. As Mono Badela wrote, in this 'non-white' world:

> the images are of dilapidated stadiums which look more like cross-country courses than playing fields. Scenes of African and coloured working class people, scrumming down on a dusty stony surface, car headlights illuminating a cold winter's night (quoted in Grundlingh 1996:195).

Why did the city simply not revamp the rugby stadium for the World Cup? Peter Alegi (2008:414) argues that image is crucial in the decision to build stadiums:

> 'The nation-state is relying on new playing facilities to market "Brand South Africa". Stadiums are seen as architectural expressions of a modern, technologically sophisticated, self-confident, proud African nation, eager to score an Eiffel Tower-like branding effect on the world stage.'

But it is more than this. The building of the stadium gave opportunity to the new aspirant black bourgeoisie to cash in on some contracts. The ANC wanted to facilitate the development of a patriotic bourgeoisie. Former President Thabo Mbeki exemplified this approach:

> 'As part of our realisation of the aim to eradicate racism in our country, we must strive to create and strengthen a black middle-class … I would like to urge, very strongly, that we abandon our embarrassment about the possibility of the emergence of successful and therefore prosperous black owners of productive property, and think and act in a manner consistent with a realistic response to the real world. As part of our continuing struggle to wipe out the legacy of racism, we must work to ensure that there emerges a black bourgeoisie, whose presence within our economy and society will be part of the deracialisation of the economy and society' (Mbeki 1999).

Geographer Peirce Lewis's summation that sporting landscapes reflect a people's 'unwitting biography, reflecting our values, our aspirations and even our fears in tangible visible form' is apposite (quoted in Nauright 1997:86).

Moses Mabhida Stadium is already a drain on the taxpayers. Conservative estimates indicate the stadium is costing the city R2 million (around US$275 000) per month to maintain. The provincial rugby outfit has decided not to move, which means that the stadium will continue to drain the city's coffers.

The stadiums standing opposite each other are a metaphor for the new South Africa. One containing memories of racial exclusion, and the other a symbol of the siphoning of public resources to create islands of privileged access in a sea of worsening inequality.

Two stadiums. Two histories. Tragedy and farce.

As one casts a gaze across the city, one gets a deep sense of how little of 'the commons' are left. It is ironic that apartheid-free South Africa has accelerated a new form of enclosure. There are already beaches where you have to pay to play. During peak seasons the beachfront is cordoned off, and only residents and those who can pay R1 200 a night to stay in hotels, are given permits. The rest have to park a distance away and walk. During international conferences, street children are spirited out of the city. Sprinklers ensure the homeless cannot find a dry spot to sleep. Police on horses bang people on their heads if they nod off on benches – just like they did when a 'non-white' sat on benches marked 'whites only'.

Still, there are places of sanctuary to be found.

Pruning at the Botanic Gardens

Nestled at the foot of Durban's Berea, a once-swanky suburb built on a ridge adjoining the city centre, is a living and breathing embellishment, 14.5 hectares in size. Standing at the highest point of this marvel next to an ancient reservoir, you see unfolding beneath you the placid undulations of trees, shrubs, flowers and grasses that are the Durban Botanic Gardens. Few others can boast such a diversity of plants: from aloes to orchids, giant fig trees to tiny lavender leaves and everything in between. Down by the lake, water birds brood, lovers edge closer, children giggle, readers flip pages, and poodles strain on their leashes. Even the lone tramp has a spring in his step.

Initially laid out in 1851 as a testing ground for new crops to be farmed in the Colony of Natal, the Berea Gardens quickly yielded a winner: sugar cane. Indeed, for many years, it was solely in support of farming that the gardens were tilled, and only in the 1890s, under the inspired curatorship of Dr J Medley Wood, were a variety of trees and shrubs cultivated. Now, with both aesthetics and biodiversity in mind, the aptly named Medley Wood went about bringing specimens from all over the world. These efforts brought high acclaim for Durban, with the gardens quickly becoming the place to be seen for picnics, weddings, and other sedate social events.

Throughout the first half of the 20th century, the Botanic Gardens continued to grow, and new features were added. During the Depression, out-of-work (white) labourers were given the task, by the city fathers, of constructing the stone-enclosed, sunken garden that is still today so pleasing to the eye. Gardening techniques progressed so that there was always some section of the parklands where trusses of flowers bloomed brilliantly, and meadows of grass were kept short and trim.

I once sat on a simple wooden bench and sucked in the late-afternoon air. Above me, the leaves of two massive exotic trees mingled, letting in shards of dark green and yellow light. The leaves were turning and, yes, I could discern the beginnings of just a little nip in the early autumn air. I had come to the park to welcome the season with an excellent bottle of Late Harvest. You see, I am decisively in the third quarter of my life. Just like the foliage above me changes colour and falls out, so does my hair. But, strangely, amidst the decay is a strange feeling of contentment and of ripeness. I don't know who else among you has experienced this curious mood.

Just for an afternoon I had vowed to stay off politics, off race, off class struggle. Alas! In talking to one of the curators, I became aware of just how encircled our existence is by such questions. The Botanic Gardens are cash-strapped because the Metro Council has pruned its public spending. After gradually being hedged into a smaller space with sprawling property developments on three of the four sides, the gardens now depend heavily on the largesse of a frugal public. There are persistent rumours that gardeners (the source of the place's creative energy) will be chopped. Some of the more frequent visitors suggest that ticket collectors and rangers should be hired to keep the riff-raff out and bring more poodles in.

With increasing costs and budgets under constant scrutiny, it is difficult to continue with the Gardens' protection of rare indigenous plants, like the African Ginger, which is clinging to survival by a tendril. Even affirmative action has to be confronted in the Gardens. With space at a premium, the curators are agonising over which trees and shrubs to cut down to make place for the admittedly under-represented indigenous flora.

Somehow though, the Gardens survive: a wonderful 'commons' enclosed on all sides by the city centre.

The Umgeni Road temple

Umgeni Road was once the main link to Durban's North Coast. With that role now taken over by a freeway, the road is home to second-hand car dealerships and a taxi rank that feeds off the huge concrete mass, which hides the city's main rail terminus. Tucked in between is one of the province's oldest temples. I passed the temple many times in my youth, but only got to visit a couple of years ago. There was a sign that greeted me. Women were not allowed to wear pants. No jeans allowed. It appeared that somebody had stolen a sign from the old Durban Club and placed it here.

Once inside, a pantheon of gods lay in wait. I walked around, trying to give the impression of studied contemplation. There was no dress code for the gods. The local priest was, himself, not exactly fully dressed. A journey had begun. Almost immediately I got lost in the world of a trinity of gods. I met Brahma, the creative principle, who emerged from the belly of another god, Vishnu, seated on a lotus leaf, and Shiva with his four hands. Shiva, who seeks to destroy the world, and Vishnu, who desires to preserve it, are perpetual rivals. Every so often, Vishnu descends to fight Shiva and to thwart the end by a few more years.

Each time he descends, he assumes a form that corresponds remarkably to the Darwinian idea of the evolution of the species. For example, he descends first in the shape of a fish, an aquatic animal; second, as an amphibious tortoise; and the fifth time as a complete human being, but short and slightly hunch-backed. The sixth time he is Parahurama, a man of great strength who lives in a forest, and whose command of language appears magical.

Frighteningly, Shiva will eventually triumph. We are living in that age now, where Shiva holds exclusive sway. This is the age of Kali Yuga – the time of destruction.

What does Vishnu do through all this? He sleeps. And we, amid the impending destruction, we must keep our dharma – the integrity of our actions. This we must do so that Vishnu does not forget the beauties of the world. Otherwise, all beauty will be erased from memory, the world will not reappear, and a new cycle will not be able to begin again. Echoes of Christianity abound. Or was it the other way round?

The triumvirate of gods is the flipside of the holy trinity: Brahma, swirling down the river on a lotus leaf and the baby Moses among the reeds; the Kali Yuga and the Revelations.

Are gods capable of plagiarism?

There is a message in the story of Kali Yuga that tells of the coming of the age of rampant materialism: shopping malls; casinos; the Lotto; the Stock Exchange … Snapping back into the present, I think of the priest worried about a dress code. I think about priests who spew out mumbo-jumbo at weddings, and the plethora of other ceremonies that masquerade under the banner of Hinduism. I think about those who still believe in the hierarchy of colour. What will they say when they realise that the meaning of the name of the eighth avatar of Vishnu (the much-revered Krishna) is black?

Visit the Umgeni Road temple. Indentured labourers found refuge there at the turn of the century. There is science fiction there, romance, betrayal, the apocalypse, psychoanalysis, and gods that are fallible. The more one reads, the more one realises that the challenge for the followers of Shiva and Vishnu is to rescue Hinduism from religion.

The present is history

Just over a decade into democracy, I ventured onto the Golden Mile, a beautiful stretch of beach whose choicest spots were reserved for whites, back in the day. As a child, my parents had brought me there on Sunday afternoons. We sat in the car and watched the white kids jump on the trampoline and whizz around in go-karts. At the 'white' cars, waiters served food and drinks on trays that attached to the windows. My father brought a flask and cheese-and-tomato sandwiches.

My mother was once briefly detained by police in the area. They thought my light-skinned father was white and that my parents were breaking the Immorality Act. In 2011, she still loves telling the story, with a hearty laugh.

At the turn of the 21st century, the 'whites only' signs are gone. The Bay of Plenty is colour-full, once again.

The year is 2001. Rickshaw puller no. 8, Johnson Zulu, is 59 years old. There is not an ounce of fat on his lithe body. This man has not seen the inside of a gym, neither has he heard of Weigh-Less. His body has been honed from years of hard work – at first on the railways, but for the last 17 years as a rickshaw puller. There might be other contending factors: He eats mostly *putu* and rice – which he cooks himself for dinner – but meat, at the most, only once a week. There might be genes at work here, too. Johnson is born 'from the same testicle as King Zwelithini'. Unlike the king, whose warrior ancestry hasn't stopped him from growing soft, Johnson works for a living. I pay for a ride. My 115-kilogram friend accompanies us. Johnson bends his back ever so slightly and we are off.

Suddenly, the rickshaw tips backwards. My friend tries to throw his huge belly forward to save us. But it has no effect – Johnson's feet are in the air. Suspended, he dangles there for a moment as if the weight is too much. And then he is back on terra firma.

The movements are imperceptible. Like the movements his forebears used to kill British soldiers with a short stabbing spear, in times past. He turns and laughs with a wide, white smile. No caps or false teeth. His feet are rooted forever in this land.

Johnson works seven days a week. Some days he makes R20 (around US$3), but during the December holidays, up to R100 and R150 a day. But things are tough. He lives in a hostel in KwaMashu, where he shares a room with four other men. He pays R50 a month for the accommodation, R21.50 a week for rail transport, and R25 a week to park his rickshaw in a local garage.

What's life like in the hostel?

> *'Most of the people are not working. We have a gentleman's agreement that we will help pay the rent of those who are not working on the understanding they will do the same when they start working.'*

Ubuntu loosely means 'a human being is a human being only through his relationship to other human beings'. Today, among other things, it is packaged and used by psychologists attached to big business to argue that workers must obey authority. The term is also used in speeches by the gravy-trainers to prove that they are human despite, their inhuman economic policies. Listening to Johnson Zulu, one realises in its truer form, *ubuntu* is honestly more alive here, among the poor.

Who are his main customers?

> *'Sotho speakers from Gauteng and Free State. They love going on the rickshaw. They are so generous. Sometimes we make so much money, we slaughter a cow. I have actually learned some Sotho. Second are the whites from Free State. That is our main business.'*

Foreign tourists hardly feature.

When does he see his wife and four children?

> *'At most, four times a year. It is expensive. R140 (around US$20) return trip.'*

How does his family receive him?

> *'When I arrive with sugar, groceries and clothing, they are excited. When I arrive with nothing, they are dull and dim.'*

On his body are centuries of dispossession. In his blood are centuries of resistance. Reduced to a number. Reduced to carrying the burden of the new South Africa on his back. A non-racial burden, mind you. Who has the gall to say South Africa has not changed?

What about the children?

> *'My greatest hope lies with my youngest of 10 years, Sibusiso. I hope he finishes schooling and becomes a clerk or a policeman.'*

When will he retire?

> *'As long as my knees are still okay, I will have to carry on. But if I win the Lotto, I will stop. I play sometimes up to R10 a week.'*

What is his wish?

> *'I want to see my family more often. I have never seen my children grow up. My wife and I have grown old separately. It breaks my heart. To be more often with my family, I wish for busloads of tourists.'*

I do not know how to tell him that he could be out of a job soon. A multi-million rand promenade squeezes the rickshaw pullers into smaller spaces, while a monorail attracts the attention of the city.

Who is his hero?

> *'When I grew up in the Royal Kraal at Nyokeni in Nongoma, we listened to many stories. Of many kings. Of all of them, my favourite is Shaka.'*

Why?

> *'During Shaka's time there were jobs.'*

Shaka is still fought over. Ten years after I spoke to Zulu, a statue of the Zulu warrior was pulled down at the spanking new King Shaka International Airport. A complaint from the Zulu Royal Family that the statue was insulting, because it cast Shaka as a 'herd boy', meant that R3 million (around US$450 000) was banished to the scrap heap.

While the provincial government seeks to build a new statue (costing another R3 million) the braves of Shaka beg on the streets of Durban. The King waits on handouts from the Provincial Treasury, while his minions demand more. Warriors dance for the very British tourists, whose ancestors destroyed them through the Hut Tax. Centuries-old routines are shortened to fit in between lunch and dessert.

If you drive down the North Coast, you will come across some beautifully structured Zulu men on the off-ramp at Ballito, handing out flyers. These flyers direct you to Zimbali where sugar estates have been turned into real estate: stolen land, stolen again. Zimbali is the acme of the seige mentality of the terrified rich. There are security guards and multiple boom-gates. It is an eerie reminder of the British garrisons in this neck of the woods in the late 19th century. If you do manage to get inside, the new white and black elite lounge at the swimming pool or play a round of golf in this wondrous *apart-topia*.

This government's economic policy, which entrenches private property, entrenches theft.

Fire on the Ridge

Every fortnight, a truly remarkable event takes place in a park overlooking the city. Just after dark, like vampires, a coven of young people are drawn together.

Parking their beat-up cars on the road, they make their way in dribs and drabs across the park, towards the shadowy glow of an ultraviolet light. There they stand in a circle for a while, saying nothing much until the rhythmic beat of ambient transcendental music finally takes up. In the gloom, one can barely make them out, holding chains and sticks and staffs in front of them.

There is the smell of paraffin, and suddenly, a flicker of fire originates. The circle slowly converges upon the single flame, which is solemnly shared. Their arms outstretched, these magical, night-time beings eagerly await their piece of flame, which comes to burn at the extremities of the implements they hold. And then the swinging and the throwing begins.

Tracing slow circles in the air, and then fast figures of eight, and then faster still and more complicated the movements become. The whole body is involved, silhouetted dramatically against the dangerous passage of fire as it inches away from eyes and hair and dress.

After a while, there is nothing but a whirling pattern of insubstantial fire in the air above heads, between legs, brushing cheeks, and searing the grass below. The fire is everywhere and nowhere, being spun and slashed and juggled all at once. And if you look long enough and surrender to the music, you see that it is the most primal of dances, with movements so elegant and light that you feel you have to join in.

I did not recognise her at first. She was young and pretty but unremarkable, barefoot in her calf-length black dress, as she stood alone on the outskirts of the ring of fire. But then, when the fury of fire-swinging was at its most intense, this majestic creature revealed herself. She waded into the centre of the circle to assume her golden throne. She was in a league of her own. She paused as a friend rushed over to light the twin chains she held in her hands, and then began the most casual of revolutions, gradually gathering not speed but subtlety.

The movement was not generated in her shoulders and hands as far as I could see. It began in her hips and passed on to the chain through that part of her arm between her elbow and wrists. That is where her body ceased and the chain began. Such grace, I have never seen before. So intricate the movements and so delicate the variations she accomplished, that I stood totally mesmerised. I doubt I would ever have emerged from this charm, had she not released me from it. With an impetuous spell of giant looping circles getting faster and faster, she suddenly sucked the life out of the flames and swung with a flurry – so fast, the fire was blown out.

There was quite a big crowd of onlookers like me, sitting around on blankets and deckchairs watching the magicians, drinking wine and eating sandwiches that were made at home. Where else in the city could public space be used, without money having to be spent?

These two-dozen kids seemed to be custodians of a carefree tribal life of drumbeats and fire-dancing, rejecting shopping malls, night clubs, and cosy nuclear families around the television. How terrible to be, I thought, part of the black and white political elite across the city on Thursday nights, squashed into suits and ties, quaffing drinks and nibbling on smoked salmon at cocktail parties, and living such fluorescently inhibited lives.

The toxic Basin

Halfway between the city centre, and what was once the Durban International Airport, is the South Basin, with the 'coloured' township of Wentworth lying at its centre.

Here we have huge petrochemical refineries that form the focal point of the province's chemical industries.

The development in the South Basin was carried out at the instigation of white business interests, with the aggressive support of both the local and national government during the days of apartheid. This meant the uprooting of whole communities in the South Basin. In Clairwood, for example, the population declined from 50 000 in 1960, to 5 000 in 1970. This 'creative destruction' was done under the code words of scientific management, implemented by planning bureaucracies at the local level.

With the new government earmarking the South Basin as a priority development zone, the continuities with the old are stark. The alternatives for the communities are also out of the dusty social engineering textbooks in City Hall: relocation – voluntary or otherwise. People living in an area known as the Barracks, in Wentworth (where rows upon rows of narrow, poorly-built dwellings give the impression of fowl runs), have been threatened with removal to a place some distance from the city centre, in dwellings so badly built and cramped and isolated, that they have been dubbed 'toilets in the veld'. For the poor in today's South Africa, development has meant moving from a fowl run to the lavatory. South African petroleum refiner Engen has started to buy up property around the plant, and is gradually industrialising greater and greater parts of the South Basin habitat.

It is not as if there is not a lighter, if farcical, side to the narrative of City Hall's supine attitude towards big business in Durban. The then Deputy Mayor, Logie Naidoo, weighed in with his own solution, which he learned from a trip to Houston, Texas: plant trees around the two major oil refineries. A local journalist retorted:

> We all like a tree, don't we? They beautify our environment, house all manner of colourful birds – and provide a vital function for male dogs everywhere… But a solution to the toxic soup which hangs dangerously over South Durban? … Excuse me, but this sounds like a smokescreen to me. Tell the oil refineries to clean up their act, perhaps. Have legislation passed that prohibits them from doing what they damn well please … I sometimes think these politicians cannot see the wood for the trees (The Mercury 2002).

But endemic and chronic asthma, at rates four times the national average, is no laughing matter. Neither is a suspiciously high incidence of blood and lung cancers. An environment conducive to the well-being of each citizen is the promise of the new

South Africa, but stunted children, choking air, and a repellent aura is the reality for those living in Wentworth.

However, Wentworth is not just a place of victims. It is a centre of the most sustained struggle against giant transnationals. Over the years, the community has taken on these giants by building local and global networks of solidarity. For those with some lung capacity to spare, one can join the irrepressible activist of the South Basin, Des D'Sa, in what is called 'the toxic tour'.

Cato Manor

In the valley just below and west of the University of KwaZulu-Natal (UKZN) is the suburb of Cato Manor, a place overflowing with history. It is here that African city dwellers found a place to live during the first half of the twentieth century.

There were riots here. In 1949 Indian landlords and shopkeepers were targets. In 1959 women led a struggle against the local apartheid state that confiscated 'illegal beer' and sought the forced removal of African residents. At the forefront of the 1959 uprising were two icons of anti-apartheid resistance: Florence Mkhize and Dorothy Nyembe.

An idea that permeated Cato Manor, according to historian Iain Edwards, was that of a 'New Africa', which encapsulated 'notions of African ethnic unity, the acceptability of revenge, and disrespect towards established notions of legality' (Edwards 1989:51).

Today, one can get a glimpse of this history in a set of evocative photographs at the Cato Manor Heritage Centre. But to get a real sense of the area, one has to take a 'reality tour' with 19-year-old Thando Manzi.

Thando is a first-year student of Geography and Environmental Management at UKZN, and on behalf of the Centre for Civil Society, he takes people around Cato Manor as part of their 'reality tour' series.

He knows the area incredibly well. He points to the river that runs through Cato Manor, and shares a memory from his childhood: when he was six years old, his friend Ntuthuko Dlamini was electrocuted and died on the spot – a young victim of 'illegal' electricity connections that passed through the river. We pass the new Reconstruction and Development Programme (RDP) houses that seem overwhelmed by shacks, some of which date to the mid-1980s. Some are 20 square metres in extent (half the size of the old apartheid-era 'matchbox' township houses).

We cross the river and enter an area that was once the SPCA. People found the smell of burning dogs unbearable. Eventually, the SPCA left, and today, people occupy the abandoned buildings and one-roomed cardboard houses.

Opposite the SPCA lies the skeleton of a biodigester. It was designed to allow the breakdown of Cato Manor residents' excrement into methane for domestic use, according to Agama Energy, by 'diverting a municipal sewer into the biogas digester, after which the supernatant flows through a series of algae and fish ponds. The final discharged water is used for food production, with the overall facility being operated by the local community co-operative'.

Though it sounds like a cutting-edge environmental project in principle, Thando described an intense community *toyi-toyi* protest against the construction team in 2009, since local residents were never properly informed about why they would be guinea pigs. They called the biodigester a 'shit fish tank', and were repelled by the idea. After spending quite a bit of money in the two years that the project was operational, the project was abandoned in 2010 as the municipality apparently couldn't come to grips with community opposition. The fence surrounding the area in which it sits has been broken, and the sloping fields have been reclaimed by informal maize growers.

The tour ends with us having a meal of tripe prepared the traditional Zulu way by Thando's grandmother.

To understand the poverty of Durban's much vaunted and award-winning post-apartheid urban planning, take a tour of Cato Manor.

Building the new (apartheid) city?

> If our urban world has been imagined and made, then it can be re-imagined and re-made. The inalienable right to the city is worth fighting for. "City air makes one free," it used to be said. The air is a bit polluted now. But it can always be cleaned up (David Harvey 2003:941).

Everyone, including the fisherfolk, street traders and African migrants, struggles to find a space in the city centre. To the south, residents fight to breathe. To the north, in the shadow of the lights of Moses Mabhida Stadium, shack dwellers fight pitched battles to defend their 'illegal' electricity connections. But largely, these are desperate and separate struggles, and in many cases the state is adept at responding in ways that divide and dissipate collective struggles. The installation of the pre-paid water and electricity metre, for example, becomes an individualising, self-policing mechanism where payment is made before delivery.

In the face of this, what will be the spark for those fighting their desperate and separate struggles to unite? What will the organisational forms of this movement do to ensure that the localised and particular struggles remain vibrant and militant? What will be the languages that unite these struggles? The answer to these questions will ultimately determine whether post-apartheid Durban will become a city that includes, rather than excludes, that opens spaces, rather than closes them down.

References

Alegi, P. 2008. 'A nation to be reckoned with': The politics of World Cup stadium construction in Cape Town and Durban, South Africa. *African Studies* 67 (3): 397–422.

Hudis, P. and K. Anderson. 2004. *The Rosa Luxemburg reader, part one: Political economy, imperialism, and non-Western societies*. Monthly Review Press, New York.

Edwards, I. 1989. Mkumbane our home: African shantytown society in Cato Manor Farm, 1946–1960. *Unpublished doctoral dissertation*, University of Natal, Durban.

Grundlingh, A. 1996. Playing for power? Rugby, Afrikaner nationalism and masculinity in South Africa, c. 1900–c. 1970. In *Making men: Rugby and masculine identity*, J. Nauright and T. Chandler, eds. Frank Cass: London.

Harvey, D. 2003. A right to the city. *International Journal of Urban and Regional Research* 27 (4): 939–941.

Mbeki, T. 1999. Speech at the Annual Conference of the Black Management Forum.

Nauright, J. 1997. *Sport, cultures and identities in South Africa.* David Philip, Cape Town.

Said, E. 2003. *Culture and imperialism*. New York: Random House.

The Mercury. 2002. Deputy Mayor can't see the wood for the trees. 30 August.

Chapter 4

Durban's Water Wars, Sewage Spills, Fish Kills and Blue Flag Beaches

Patrick Bond

Water is one of the primary barometers of climate change: A rise in sea-levels, flooding, and extreme storms combined with general water stress and more severe and frequent droughts will escalate crises in municipal infrastructure, requiring continual upgrades for water purification, stormwater drainage, and sewage treatment, all of which will dramatically raise the price of water at the retail level. In South Africa, the dry western side will be most adversely affected by droughts (threatening the production of rooibos tea and Cape wines). According to the Academy of Science in South Africa (ASSAf), Durban is also at great risk and will experience higher temperatures and heat stress, volatile rainfall, up to 160 million cubic metres less water each year by 2100, a sea-level rise of up to a metre by 2100 across Durban's 100 km of developed coastline, lower biodiversity, higher disease levels (especially malaria and cholera), declining agricultural output (a one degree Celsius rise leaves the surrounding region unreliable for the staple maize production), and other economic stresses (ASSAf 2011:27).

Tourism, one of Durban's main economic engines, will be irreparably harmed. Swimmers and surfers think of Durban's beachfront as one of the world's finest in any urban context. After apartheid-era rules that prohibited black people from using the best beaches were lifted at the end of the 1980s, the area stretching from the Blue Lagoon's Umgeni River to South Beach's uShaka Marine World – including the immensely popular North Beach area near the main restaurant strip – represented one of South Africa's most impressive, open and democratic public spaces.

The beach area hosts a number of Durban's key attractions. Five blocks west of North Beach is the International Convention Centre and Durban Exhibition Centre, where the 17th Conference of the Parties (COP 17) to the United Nations Framework Convention on Climate Change (UNFCCC) is holding its 2011 summit. Durban harbour, further south, is also a major tourist site and is bordered on the southeast by the middle-class suburb of The Bluff.

Two major rivers, the Umlazi and the Umhlatuzana, cut through the South Durban Basin, emptying polluted water into the harbour and Indian Ocean. Between The Point, an area at the harbour entrance in the midst of luxury development, and the old, now closed Durban International Airport are Ansteys Beach and Brighton Beach, which are also beloved by surfers, sea-kayakers, swimmers, and fisherfolk. In a variety of ways – especially with regard to water quality – the looming prospect of substantial anthropomorphic climate change threatens these beaches and adjacent areas. A half-metre rise by 2100 – predicted by all major studies – will, as ASSAf concludes (2011:27), 'severely erode beaches and have an impact on tourism'.

In mid-2011, the national environment official Razeena Omar – who is responsible for coastal management – made the same observation about climate damage to the beaches. Acknowledging that Durban was leading South Africa in 'determining new development setback lines to mitigate against the effects of future sea-level rise and coastal erosion', she stated the following in a letter to municipal consultants regarding proposed beach-front development:

> [T]he same municipality now appears to ignore their own setback lines and is proposing placing new infrastructure seawards of this setback line. Integrated Coastal Management finds this unacceptable and trusts that the province will give this issue careful consideration before an environmental record of decision is issued. In the short term, infrastructure should be placed as far back from the high-water mark as possible, while in the long term, a controlled retreat should be considered for sections of this coastline (The Mercury 2011).

In the short term, water quality at the beach is *already* a major concern, as evidenced in recent years by the increasing numbers of surfers and bathers emerging from Durban's ocean with runny stomachs and ear infections. On many a hot day, *E. coli* counts soar, especially during periods of sand dredging or after heavy rains, which wash polluted water and silt into the city's four major rivers – the Umgeni (which supplies the bulk of the city's piped water via the Inanda Dam), Umbilo, Umhlatuzana (into the port) and the Umlazi between the two main refineries in South Durban – and then into the sea. The 2010 *State of South Africa's Rivers* report revealed that each river has 'poor' water quality at the point where it courses through the vast informal settlements on the outskirts of the city, beyond the 'sanitation belt'. Geographically, that belt represents the distance and housing density that deters municipal planners from installing water-borne sanitation – although the numerous exceptions make it clear that socio-economic relations rather than spatial relations are the decisive factor.

Perhaps the so-called sanitation belt should be tightened up several notches to include the wealthy, mainly white areas of Durban. After all, it would be ideal for the *entire* society to switch from water-borne sanitation to a system that uses biodigestion to break down excrement into methane for home utilisation and creates pathogen-free fertilisers, utilising the phosphorus that can be extracted from urine. Engrained class and racial privilege means these neighbourhoods will be the last to convert from water-borne to dry sanitation. But the expense of building and maintaining such biodigestion systems was illustrated in the failure of the main biodigester pilot project in the black community of Cato Manor, where the 2010 municipal retraction of subsidies doomed this laudable experiment. Fortunately, a similar one is 'operational in Cato Crest and yields 60kW of electricity for use at the community centre', ASSAf (2011:56) reports. However, this project is not yet widely known in Durban.

Instead of this more collective approach – which would turn methane into energy for community use – a different sanitation system based on short-term, atomised, money-saving technology was imposed on low-income people: Urine-diverting toilets[1], which proved unpopular in many of their first 90 000 installations Behind that top-down decision, which precludes an entire area's ability to incrementally upgrade to the sanitation grid, lies another: pricing the consumption of sufficient water (e.g., up to 12 litres for a standard toilet flush) beyond the realm of affordability for a substantial – perhaps majority – share of the Durban population.

But none of this should obscure the extraordinary attention that Durban has received in the water sector for the supposed competence and entrepreneurialism of its senior management. Thus, the critique will not focus on the personalities and foibles of the key players (though it is impossible not to record some idiosyncrasies) but instead will examine the *policies* and *political economy* of water provision. For it will take a profound structural change to ensure climate change adaptation does not lead to even worse forms of water apartheid than currently exist. This is especially true because the authorities are pleased with their policies, programmes, and projects.

Durban's Orwellian water awards

South Africa's second largest metropolitan area, Durban (or eThekwini) has an exceptional reputation for water and sanitation delivery. One academic proponent of Durban water management, Zoe Wilson (2006:14) noted the city water department's:

- National and international acclaim
- professional acclaim for the head of eThekwini Water and Sanitation
- financial solvency and well-maintained infrastructure
- overall customer satisfaction
- strong rural service track record
- emphasis on sustainability and alternative technologies
- emphasis on job creation, and
- inter-departmental umbrage.

One of the world's most prestigious periodicals, *Science* magazine, featured eThekwini Water and Sanitation head Neil Macleod in a 2008 article:

> Macleod's department confronted the challenge of rapidly expanding water and sanitation services in 'township' settlements while keeping its budget afloat. At the time, a quarter of a million households in the Durban area had no access to clean water or sanitation. To jump-start improvements, Macleod got permission from the city in 1996 to provide a daily 200-litre water ration – a policy that became a national goal (Koenig 2008:744).

The impression soon emerged that Durban hosted one of Africa's finest water services departments. Macleod was recognised with a string of awards: the 2002 *National Geographic* award for sanitation, the 2003 Dubai International Award for Best Practices (for sewage disposal education), the 2003 South African national 'Excellence in Innovation' award, the Ford Foundation's 2003 Impumelelo Award, and the 'best municipal delivery' and 'Blue Drop' awards in the last half of the 2000s. As Bill Gates (2010) blogged, Macleod 'has been a leader in thinking through how to improve sanitation for the poor in Durban'.

These claims of success, however, call out for more careful professional and academic scrutiny than Durban has so far been given – especially if the municipality continues to serve as the country's pilot for national water policy and its bulk water infrastructure, distribution, retail supply, and disposal practices allegedly set the standard for South African and African cities.

If on the contrary, the city's water system is breaking down – as the Durban press and concerned citizens have been recording – then a systematic analysis of the causes is crucial. This is especially true if Durban's water woes relate to broader neoliberal policies and practices, including the recent mal-pricing of water and artificially imposed fiscal austerity. These shortcomings will, after all, amplify with climate change, when yet more stress is placed on decaying infrastructure from extreme flooding and rising seawater, and as other exogenous factors, such as high pollution levels (from mercury to *E. coli*) impact the water catchment system as a result of increased coal-burning and inadequate upstream sanitation.

Likewise, on the household retail side, Macleod promised that by 2010 'everyone' in Durban would have basic water and sanitation (Koenig 2008:744). In 2011, it is fair to estimate that many hundreds of thousands of residents of shack settlements still do not. A decade ago, Durban was posting relatively mediocre statistics for water and sanitation access in relation to peer cities. Indeed, surveys by the SA Cities Alliance, a network associated with the World Bank, reveal that Durban led South Africa's major cities in the *decline* in the supply ratio of (in-house) water to residents during the early 2000s.

In 1996, 62 percent of Durban's residents had water within their household, but by 2004 the rate was reduced to 50 percent, mainly due to municipal boundary changes. The growth in population without adequate sanitation rose from 30 percent in 2001 to 32 percent in 2004 (SA Cities Alliance 2006:3). Even as recently as 2007, municipally sponsored water-customer focus groups in three representative working-class and poor areas identified unaffordability (by more than 75 percent of those surveyed) and city-society conflict (between 20 and 50 percent of those surveyed) as the main reasons that those surveyed lacked water and sanitation services (Wilson, Malakoana and Gounden 2008:142–146).

To be fair, the task of providing municipal water delivery was and remains extremely difficult. In 2001, what were 840 very diverse and fragmented municipalities were reduced to a 'wall-to-wall' coverage of 284 new local governments, including six metropolitan areas. The percentage of Durban's low-income inhabitants without direct water supplies soared simply due to the new boundary demarcation.

Meanwhile, government's national-local financing flows fell short in the wake of substantial real budget cuts during the 1990s (Bond 2000) and inadequate 'Equitable Share' funding[2] available during the 2000s. Structurally, the society was experiencing a dramatic increase in unemployment during the late 1990s, doubling from 1994 levels of 16 percent to 32 percent, while the Gini coefficient measuring income inequality continued to rise steadily from 1994 throughout the 2000s (Bond 2010b). Over the entire post-apartheid period, according to the SA Labour and Development Research Unit in 2010, 'poverty incidence barely changed in rural areas, while it increased in urban areas' (Leibbrandt et al 2010:36).

Instead of making much more substantial investments in water infrastructure and the regulation of water pollution, in 1994 the national government turned more explicitly neoliberal. Strongly influenced by the World Bank and neoliberal donor agencies (Bond 2005; 2006), the Department of Water Affairs and Forestry had adopted a 'width not depth' approach to many basic need services, including housing, water and sanitation, and electricity, which adversely affected both the retail-level supply and related bulk infrastructure (Bond 2002; McDonald and Pape 2002; McDonald and Ruiters 2004). The roll-out of water services proceeded with quantity (number of communities connected to collective taps), not quality or sustainability in mind. The percentage of GDP devoted to water, gas, and electricity actually declined after 1994, and the number of municipal engineers dropped by half (Bond 2005; 2010b).

Locally, controversies emerged when civil society challenged Durban water officials over various aspects of water delivery. Macleod made concessions that do indeed provide some optimism for longer-term trends in access, management, and water governance. Just as the 1973 dockworker strikes in Durban's harbour signalled the beginning of a major trade union movement that became the continent's strongest and a decisive factor in ending apartheid, the 1999 mobilisation of thousands of the city's Chatsworth neighbourhood residents – bringing poor and working-class people of Indian and Zulu ethnicity together in rare combination – heralded not only the new post-apartheid urban social movements, but also an intense set of battles over water access and pricing, leading to widespread illegal reconnection of water following mass disconnections.

It should be noted, however, that despite the implications for those who cannot pay their water bills, ASSAf and Durban authorities consider illegal connections – which contribute to a 40 percent rate of unaccounted-for-water – as an opportunity to mitigate greenhouse gas emissions (to a hoped for 23 percent by 2016) by disconnecting the unlawful hook-ups.

> Water losses in the distribution of potable water have a high environmental burden in addition to financial implications. This should be the first and most important step in reducing [g]reen[h]ouse [g]as emissions. Taking

> into account the high proportion of unaccounted water (e.g., due to leakage, illegal connections, etc), reducing supplier demand will have the highest environmental and financial improvement impact without negatively affecting the quality of the supply (ASSAf 2011:110).

ASSAf (2011:55) also commended Durban officials for 'rectification of unmetered connections and informal settlement metering', as well as 'new supply pressure zones to reduce average water pressure' even though the amount of pressure is a crucial factor in tap water's ability to eradicate water-borne disease carriers – a major factor in KwaZulu-Natal's public health disease burden. Meanwhile, although many low-income community organisations fought back against higher water prices – e.g., July 2011 saw a 39 percent increase at a time when the inflation rate was less than 6 percent – by encouraging further illegal reconnections, others were partially co-opted into advocating that communities reduce their water consumption along the lines of the 'Citizens' Voice' strategy. For example, one local NGO, Mvula Trust, aimed to raise awareness about poor people's alleged need to lower their water consumption:

'"Citizens' Voice" contributes, through extensive public education, to greater household awareness in conserving water' (Water Information Network South Africa 2009). The strategy is implemented in areas that are preselected by the municipality, based on where high levels of water losses have been recorded.

This objective – saving water by denying it to low-income people through various technologies, as well as a climate narrative that blames low-income victims – was made explicit in the city's Urinary Diversion sanitation strategy. But notwithstanding some concessions in 2008, including a higher level of Free Basic Water for low-income residents (Bond and Naidoo 2008), contradictions and social strife remained acute. On the retail side, this was a function of disconnections, pricing debates and service-level controversies relating to household water and sanitation. On the supply (and disposal) side, it was a result of the city's failure to invest in adequately and maintain the water services infrastructure, especially bulk supply, sewage piping, and stormwater drainage.

Sewage spills and fish kills

Wastewater contamination of central Durban's beaches is one of the primary adaptation threats posed by climate change, since, as ASSAf (2011:28) acknowledges, 'the city's wastewater infrastructure will be impacted by sea-level rise due to [its] low elevation and proximity to the coast'. It is already such a serious problem that, in a national scandal beginning in 2008, Durban's 'Blue Flag' beach status was retracted as a result of high *E. coli* counts, a matter still not resolved ahead of the COP 17 and subsequent December 2011 summer holidays. By the end of 2007 it was suddenly evident that Durban municipal officials had deprioritised water health, neglected pipe maintenance, and redirected needed funds away from sewage station infrastructure. Formal sanitation was nearly nonexistent in the low-income settlements that emerged over the prior fifteen years, with a typical shack settlement of several thousand residents often serviced with only a handful of chemical toilets.

The catalyst for widespread civic concern and embarrassing press reports was the sudden appearance of tens of thousands of dead fish in Durban's harbour during the December 2007 holidays. Concern peaked when the city's once proudly acclaimed Blue Flag beach status was revoked in 2008. According to award-winning environmental journalist Tony Carnie (2008a) of *The Mercury* newspaper, nearly two months passed before it was determined that the failure of a city sewage works was responsible for the 2007 fish kills, in part because of municipal obstruction of information:

> The findings by the Council for Scientific and Industrial Research (CSIR) follow repeated denials by municipal manager Mike Sutcliffe that raw sewage spills were the primary cause of the massive die-off of fish and other marine life just after Christmas Day. The CSIR report, compiled by researchers Steven Weerts and Shamilla Pillay on behalf of the National Ports Authority, contradicts Sutcliffe's assertion that a combination of rainfall, industrial spillages and other factors played a role in the death of thousands of fish in Durban Harbour from December 26 to December 27.

This was not the first time that the Umhlatuzana River, which drains into the harbour, suffered extreme pollution. However, on no prior occasion in living memory was the damage so severe. According to Carnie (2008b), a March—August 2007 sewage spill of 17 million litres into the Umhlatuzana occurred as a result of a major sewerage pipe blockage near Queensburgh. This was blamed on illegal dumping by a local company (Commercial Waste Services), and led to the municipality filing a lawsuit seeking damages of R1.4 million (around US$200 000). Macleod testified that because of the blockage, 'a section of the sewer main had to be re-laid and sewage discharged to the [Umhlatuzana] river for a period' – while the firm's chief executive denied the charges, accusing Macleod of 'barking up the wrong tree in an attempt to find a scapegoat for recent fish kills in Durban harbour'. Carnie (2008b) requested a corroborating report about the fish kills for the city by water quality consultants Tickey Forbes and Nicolette Demetriades (from Marine and Estuarine Research), but

> deputy health head Siva Chetty has refused to release it to *The Mercury*, despite repeated requests. Chetty was appointed by Sutcliffe to deal with all media queries on the fish kill, and his refusal to release the report has fuelled suspicions of a city hall cover-up or a reluctance to acknowledge sewage management problems.

A few weeks later, Carnie (2008g) alleged that in providing only partial reports about the broken pipes to the City Council,

> Sutcliffe appears to have gone out of his way to discredit or suppress much of the evidence … When *The Mercury* asked for a copy of [Durban city official Andrew] Mather's report, Chetty maintained that it was an 'internal report' and that some of the contents had been incorporated into the reports submitted to councillors. However, *The Mercury* has obtained a copy of Mather's report, from other sources, which says: 'There is no doubt in my mind that the fish kill is directly attributable to the amount of nutrients built up from sewerage inflow that entered the bay, and no other causes'. It also pours cold water on several of Sutcliffe's statements on other possible causes of the fish kill such as heavy rainfall washing pollutants into the bay, toxic spills or other waste-water discharges.

Another report on Durban's rivers by water quality consultant Mark Graham emerged at the same time as part of the national Department of Water Affairs and Environment's *State of South Africa's Rivers Report* on the Umgeni and neighbouring river catchments. Carnie (2008f) reported that Graham 'highlighted several cases of broken sewage pipes and overflows and poor treatment at several eThekwini municipal sewage works'. As Carnie (2008j) wrote,

> The eThekwini Municipality has been singled out as one of the 'most significant' polluters of some rivers because of the failure to repair burst sewer

lines and poor management at some wastewater treatment plants ... Where the Mlazi River flows into the concrete Umlaas Canal section, he found 'geysers' of raw sewage bubbling up from the bottom and test samples showed readings of 280 000 counts of *E. coli* [colony forming units, or CFUs]/100ml of water during his latest 2007 survey. Graham reported similar high readings of *E. coli* to authorities during the 2006 survey, but it appeared that 'little had been done' to fix the problem. In the Isipingo River, *E. coli* concentrations of 440 000 counts/100ml were recorded in the latest survey. ... [t]he South African guideline for acceptable recreational contact was set at 130/100ml.

Macleod confirmed that these problems were associated with irregular maintenance, but he also shifted the blame in a manner that can be read in race or class terms:

> We are often called in to unblock the same sewer lines. Then we find anything from sheep heads, rocks, shoes and buckets, rags or newspapers dumped in these pipes. This is a process of education, to get people to realise that sewers are not rubbish dumps (Carnie 2008j).

Graham's report, however, highlighted areas where Macleod's own department was most responsible. For example,

> the Umbilo municipal wastewater treatment works was having a "highly significant and negative impact" on the health of the Umbilo River. It was 'probably the biggest single point-source impact on the Umbilo.' Little appears to have changed in this system for the last 10 years. Parts of the Umkhumbane (near its confluence with the Umbilo) was home to rat-tailed maggots as well as "sewage/blood worms". These maggots have special "snorkels" to breathe and survive in the most highly polluted environments. Sewage worms are also specially adapted to surviving in very foul water. The Umhlatuzana, Hammarsdale and Izimbokodweni wastewater works or pump stations were all especially problematic (Carnie 2008j.).

Quoting from Graham's report (South African Department of Water Affairs and Environment 2008):

> The following are some recent *E. coli* results illustrating that the provision of goods and services by these rivers have been compromised by poor management of human activities and infrastructure relating to faecal contamination: Mayville Stream, a small stream between Mayville and Westridge Tennis Stadium, *E. coli* up to 290 000 per 100ml of water (Cause: Informal communities – inadequate sanitation). Bellair Stream, *E. coli* up to 310 000 (Cause: informal community). Aller River, New Germany. *E. coli* up to 400 000 (Cause: informal community – inadequate sanitation). Umhlangaan River above Kwamashu wastewater works. *E. coli* up to 240 000 (Cause: broken sewers in Kwamashu Township). Isipingo River above the Isipingo wastewater treatment works, *E. coli* count up to 10 000 000 000 000 and exceeding 100 000 000, 60 percent of the time (Cause: broken sewers in uMlazi Township). uMngeni River at Kennedy Road, *E. coli* up to 1 080 000 (Cause: Informal Community on the banks of the Palmiet River). This is an example of the impact that a community on a small river can have on a larger river! uMlazi River below Fongozi Stream, *E. coli* up to 720 000 (Cause: broken sewers in uMlazi township). The Slangspruit just above Pietermaritzburg and the Baynespruit within the City boundaries, have recently contained *E. coli* counts of 670 000 and 610 000 per 100 ml respectively.

What the 'poor management of human activities and infrastructure relating to faecal contamination' explicitly suggests, is the failure of Durban authorities to offer adequate sanitation in shack settlements ('informal communities'), which in turn creates vast public health threats across a very large geographical area. At the level of basic pipe maintenance, as Carnie (2008h) reported, it transpired that

> heavily polluted water from a punctured sewage pipe appears to have been flowing into Durban harbour via the Umhlatuzana River for up to nine months – reportedly the result of a bungled repair job by eThekwini council contract workers. A *Mercury* investigation on Tuesday also disclosed evidence of illegal discharges of a resinous waste into council sewers in the Seaview area. Sewage was still gushing into the river on Tuesday morning, despite the fact that municipal manager Mike Sutcliffe said on December 28 he would "leave no stone unturned" to get to the bottom of a massive fish kill in Durban harbour over the Christmas holidays…

According to staff at a nearby shoe factory, the overflow into the Umhlatuzana started in April last year when eThekwini council contract workers cut a hole in the pipe with oxy-acetylene torches to relieve pressure from a blocked sewage pipe near the corner of Umhlatuzana and Harden roads in Seaview. Neil Macleod, head of the eThekwini Water and Waste department, expressed surprise on the Tuesday in question when *The Mercury* alerted him to the problem. Macleod said he had asked staff for a full report on the matter, but he disputed that the pipe had been leaking for nine months. He suggested the problem could be very recent. Nevertheless, correspondence shown to *The Mercury* suggests that the municipal manager's office had been made aware of the blockage – and the possibility of an illegal stormwater-to-sewer-pipe connection – eight months ago.

According to Macleod, 'contractors had been employed to fix the problem between March and August [2007, but] sewage had flowed into the Umhlatuzana for approximately five months as there was nowhere else to divert the flow while the blockage was cleared' (Carnie 2008h). As for the leak in early 2008,

> 'This latest overflow into the river appears to be quite recent. It may be a result of vandalism, but I can't say that categorically until we have investigated further' (Carnie 2008h).

In February 2008, Macleod (2008b) formally replied to the charges of mismanagement of rural sanitation as follows:

> 'A piped sewerage system is not economically justifiable in rural areas, where the densities are too low, and in these areas onsite sanitation is the only viable option available. The rapid densification of the municipality has led to the run-off of untreated sewerage [sic] and polluted storm water into a number of rivers. It can be reasonably expected that the housing backlog will be eradicated by 2015 and only then will the problems faced through informal settlement pollution washing into the streams be finally solved.'

This garbled reply contains many contradictions. If population densities in rural areas are too low to justify constructing a piped sewerage system, then 'rapid densification' should be the basis for running *subsidised* sewage pipes even to informal settlements. The systematic lack of sewage and water pipes in Durban's shack settlements, home to more than 800 000 residents, should have disqualified Durban from any municipal awards. As a matter of public policy, only a handful of standpipes and chemical toilets were provided in the informal areas of Durban even if many thousands of people were resident. Macleod puts the onus for sanitation on housing construction, despite the fact that the municipality's housing department has been plagued with widespread scandals over improperly constructed residences by developers with close political connections to the ruling party.

The municipality was also cutting back its annual housing output by 40 percent at the time of the World Cup. But Macleod didn't only point the finger at low-income people for not having access to sanitation and companies that illegally dumped waste in the sewerage system, he also blamed Durban's higher-income citizens:

> Residential property owners have surfaced large areas of their property and often find that the run-off during times of heavy rainfall cannot be catered for by their stormwater drainage system. This problem is worsening as storm intensities increase. The solution used by residents is to lift sewer manholes and allow the storm water plus the attendant sand and refuse to be carried into the sewers where it is deposited and later causes blockages… In certain communities, children also see the sewerage system as a source of entertainment. Objects such as stones, twigs and glass when dropped into a sewer manhole are carried away by the effluent. What is not realised is that these foreign bodies accumulate at a change in grade and cause a blockage of the sewerage system (Macleod 2008b).

Instead of upgrading stormwater drainage capacity, Macleod chose to spend more money on citizen education. He effectively conceded defeat in denying the possibility of 'pristine' river quality (a standard no one would reasonably expect, instead of the reasonable demand to meet national guidelines for ecologically healthy rivers), in creating a 'tension' between people and aquatic life, and in positing different standards between inland and coastal rivers:

> It cannot be expected, however, that the rivers in our fast-growing municipality, at the levels of poverty and abuse of the sewerage system we experience, will return to a pristine condition. There is also a tension between achieving a river quality that is acceptable for humans and one that is acceptable for the aquatic creatures that live in our rivers. … The Graham report often refers to nutrient enrichment and nutrient levels that exceed national standards.

> While these standards may be applicable to certain inland rivers, they are not generally applicable to coastal rivers (Macleod 2008b).

In March 2008, yet another fish kill occurred:

> Hundreds of dead fish are floating in the Joyner Canal in Prospecton, Durban. And nearby, authorities are oxygenating the waters of Isipingo lagoon to try to save fish and crabs from dying … After Tuesday's downpour, the canal had flooded its banks, spilling into some adjoining properties (Ryan 2008).

Macleod admitted another factor that played a role – Eskom's periodic load shedding (otherwise known as rolling blackouts): 'There are also sewage spills into the harbour waters from time to time, particularly with the introduction of load shedding, which has contributed to the pollution load in the harbour sediments' (Macleod 2008b). Specifically, he noted, Durban has

> 292 sewage pump stations and 132 water pump stations. Of these, about 40 are equipped with standby generators – these are the stations where electricity supply was not considered sufficiently reliable at the time of installation. Load shedding has changed the picture completely. We estimate that in any cycle of load shedding about 20 sewage stations overflow to varying degrees … We monitor the condition of the rivers and have not yet seen a significant impact on the rivers in the municipality as a result of these spills – presumably because most are of short duration and comparatively low sewage volumes are involved (Macleod 2008b).

Eskom stopped load shedding after mid-2008, when metals and minerals prices crashed. This reduced demand for electricity in smelters and mines and resulted in a reserve of electric capacity. No sewage spills due to load shedding have been reported since. However, Eskom's two new coal-fired power plants – the world's third and fourth largest, once they come on line beginning in 2012 – will dramatically increase electricity generation capacity, which will in turn contribute to South Africa's exceptionally high carbon dioxide emissions. Whether the additional electricity drives enough new demand to result in future electric grid outages' remains to be seen.

As mentioned above, climate change and extreme weather events are expected to exacerbate Durban's water management challenges. In March 2008, for example, heavy rainstorms led to the closure of the two major Durban oil refineries, and according to Carnie (2008d),

> Municipal staff also had to break open the mouth of the Isipingo River yesterday to prevent a tide of raw sewage from wiping out fish life in the Isipingo estuary after several large sewage lines were torn open. Some sources estimated that up to 30 million litres of untreated sewage had flowed into the river in a day, and it could take several days to plug the flow… The storm caused severe flooding of homes, businesses and other buildings and several major routes could not be used because of high water levels… One of the immediate consequences for neighbouring residents in Wentworth, Austerville, Merebank and other suburbs were clouds of burning petroleum vapours being emitted from the flaring stacks of both refineries. Desmond D'Sa, of the South Durban Community Environmental Alliance, said residents had been subjected to 16 hours of flaring from the Engen refinery. Flaring is normally an emergency process whereby the build-up of high levels of petrol vapours are burned up rapidly to reduce the danger of explosion during a sudden shut-down. Elsewhere, the floods ruptured a large sewerage pipeline in at least two places in the Umlazi area.

Even before climate change effects are fully felt, the damage already done to Durban because of inadequate maintenance on pipes, stormwater drainage under-capacity, and sewage station breakdowns has been severe. Several beaches are the core attraction behind the city's leading position in domestic tourism, yet when six Durban swimming beaches lost their international Blue Flag status between 2007 and mid-2008, the reputation of these sources of urban pride seriously declined. The initial reasons for the downgraded status of two main beaches, North and Addington, were 'poor quality of ablution facilities, litter problems and poor beach sand quality'. But Carnie (2008e) subsequently reported that

> regular tests showed the two beaches had been unable to comply with World Health Organisation and South African standards for recreational water quality… 100 parts of faecal pollution/100ml of water, whereas Durban samples collected on February 12 [2008] showed readings as high as 960/100ml.

These findings were immediately dismissed by Sutcliffe:

> We have a more realistic view now of what we want to achieve, and I am not overly worried. Some issues are dependent on storm water, and unfortunately these systems were designed well before the beaches were there and the outlets now flow on to those same beaches (DeBoer 2008).

Blue Flags down

The reaction to the very high *E. coli* counts by those closer to the scene, including the city's project executive for coastal and catchment policy, Andrew Mather, was panic: 'We are trying to find out what the problem is as quickly as possible, and then rehabilitate our Blue Flag programme' (Carnie 2008i). Once it was determined how difficult it would be to solve long-standing problems of inadequate maintenance, Sutcliffe and Macleod argued against the Blue Flag criteria on grounds of inconsistent bacterial measurements between different temperature zones. Sutcliffe also claimed (without specific evidence other than hearsay) that Blue Flag South African coordinator, Allison Kelly, was 'involved with some oppositional groups who have their own agendas to portray our city poorly. Kelly thinks the developing world must have different standards applied to our beaches' (Christianson 2008). Sutcliffe's attempt to have her fired from the Blue Flag programme (bragged about in Sutcliffe and Macleod 2008) was rebuffed by the international Blue Flag headquarters in Copenhagen, which declared her work 'outstanding' (Carnie 2008k).

Writing together (or so it was advertised on the municipal website), Sutcliffe and 'McLeod' (sic) made the following hyperbolic claims:

> The recent politically inspired and charged media statements which have surrounded the cleanliness of our beaches require comment. Instead of relying on facts and scientific results, media reports have sensationalised comments made by a self-proclaimed spokesperson Alison [sic] Kelly in which she has not only ignored pollution data and our own analyses, but interestingly, she says nothing about the fact that in Europe the so-called Blue Flag status is not as strictly enforced as in our country. Colleagues have shown me photos of Blue Flag beaches in Europe where animals freely roam and defecate on the beaches …

> In the past year, only two beaches have exceeded the 2 000/100 ml limit that requires immediate withdrawal of [B]lue [F]lag status. The other beaches show generally consistent, good quality water. The [B]lue [F]lag standard permits 20 percent of samples taken, to be above 100/100ml and Ph [sic] and other parameters are also within the limits set. Clearly not happy with such results Kelly now complains that she wants an independent laboratory to undertake the studies (Sutcliffe and Macleod 2008).

Yet as for the admitted problem of stormwater drainage pollution contaminating the city's main beaches, Sutcliffe and Macleod (2008) blamed

> the behaviour of formal and informal traders and street dwellers which lead to pollution of these drains. Poor water quality bacteriologically is not always as a result of a failure of the sewerage system per se but of general misuse of the system and poor behaviour in catchments. The likelihood of this being the source of many of the problems seen is supported by the fact that our water quality failures are most often as a result of high Enterococcus levels – an indication that the pollution event is removed either in time or distance from the sample.

Sutcliffe and Macleod (2008) ultimately denied that there was anything to worry about:

> Our beaches are packed in spite of Ms. Kelly and others doing their best to portray us in a poor light. We will continue to work to ensure our beaches remain the cleanest and safest compared with anywhere in our country.

There was, in this vitriol, only a hint that the city's own infrastructure lacked the physical investment required to adequately dispose of both sewage and stormwater in ways that didn't threaten public health. The failure to retain Blue Flag designation on the six beaches was one crucial indicator of this threat, though Sutcliffe and Macleod both continued to claim otherwise and mischaracterise the Blue Flag standard. For example, Heinz de Boer (2008) reported in *The Daily News* that

> the chance of local beaches ever flying another blue flag seem very remote, city manager Michael Sutcliffe said. 'We are more advanced than the Blue Flag anyway because their approach has been inconsistent and not developmental in its approach. The main issue was never water

> quality, but the exclusion of some of the most beautiful beaches in South Africa simply because they may not have had recycling bins or other facilities available'.

In early 2010, a city council meeting unanimously agreed to an opposition (Democratic Alliance) councillor's proposal that Durban again apply for Blue Flag status at the six beaches. This followed a substantial critique of Sutcliffe from within his own party: 'Tourism Minister Marthinus van Schalkwyk and then tourism [provincial minister] Weziwe Thusi had urged the eThekwini management to reapply for the status, saying it was important as a tourism drawcard' (Padayachee 2010). But in August 2010, opposition council member Geoff Pullan (2010) wrote, pessimistically, that the city's proposed September 2010 application would not succeed:

> Unfortunately, the record will show that eThekwini, which had Blue Flag beaches from 2001 to 2008, has not been able to regain its high level of seawater quality. Not being able to improve the quality of water run-off into the sea, will be the legacy that the present administration will be remembered for.

In mid-2011, Macleod's department issued a water quality report on all 175 rivers and tributaries in the Durban metropolitan area, which found 90 of them 'near natural', 'good', or 'fair'. However, this report neglected to test for *E. coli*. Instead, 'those involved in the programme studied the state of living organisms in the water' (SAPA 2011). The monthly updates on Durban's official website show no overall progress and that erratic measurements mar the monitoring system. Moreover, there is a race-class-geographical factor that has not been officially acknowledged but that is discernible in reports of beach water quality on certain days. For example, in late August 2011, the department reported 'excellent' water quality from central Durban northwards, 'moderate' quality at the point where the (mainly white) Bluff area moves southwards from largely upper-class to middle-class neighbourhoods, and 'poor' quality where black South Africans swim at Treasure Beach southwards, and where the city has placed its famed shark nets to prevent attacks (eThekwini Municipality 2011).

Because of the overall failure of Durban's sanitation programme (Bond 2011), tests of *E. coli* are critical. As if to emphasise the point for the sake of the international press, an hour south of Durban – in the Margate municipality's Mpenjati Nature Reserve – a world-famous journalist, Bill Johnson, lost part of his leg (and nearly perished) in March 1999 due to flesh-eating bacteria caused by the accumulation of *E. coli* in a stagnant lagoon near his holiday cottage. As he recorded in the *London Review of Books*,

> Almost certainly the reason the lagoon was polluted with such a deadly organism was to do with the dumping of raw sewage by communities living upriver. Only months later was I able to Google necrotising fasciitis and [found] a long list of famous people who died from the disease, usually within 24 or 48 hours of contracting it (Johnson 2009).

According to a *Daily News* report, '14 strains of bacteria were found in Johnson's wound', one of which was becoming a serious threat in the greater Durban region:

> Vibrio vulnificus, of the same genus as that which causes cholera. The bacteria species occurs in warm, salty water, usually in sub-tropical and tropical areas. However, with global warming, reports of fatalities from Europe are beginning to be recorded. According to the United States Centers for Disease Control (CDC), Vibrio vulnificus can cause disease in those who eat contaminated seafood or who have an open wound that has been exposed to seawater. In healthy people, ingestion of Vibrio vulnificus can cause vomiting, diarrhoea and abdominal pain. In people with compromised immune systems, particularly those with chronic liver disease, Vibrio vulnificus can infect the bloodstream, causing severe and life-threatening illness characterised by fever and chills, decreased blood pressure (septic shock) and blistering skin lesions, the CDC says. It can cause a skin infection that may lead to skin breakdown and ulceration (Maluleka 2009).

It was soon obvious that this was a generic problem, with *The Mercury* reporting that other communities south of Durban were also

> fighting an uphill battle to prove that local swimming beaches are heavily polluted with sewage from local rivers and storm water drains. Frustrated by the apparent failure of officials to control sewage pollution around beaches and lagoons, they have gone to the extraordinary length of collecting their own water samples and getting them tested at an accredited laboratory in

> Pietermaritzburg. The association has just released copies of the analysis, which suggest that some rivers are severely polluted with *E. coli*, a bacterial indicator of sewage pollution from humans or animals.

According to the report from Pietermaritzburg-based Talbot and Talbot Laboratories, samples collected from the Makamati River contained *E. coli* levels of 28 800 units per 100ml of water – several thousand times higher than the Department of Water Affairs' guideline levels for safe recreation in fresh water. The guidelines suggest that there is a risk of people developing stomach infections and other water-related infectious diseases, even at exposure levels as low as 130 units of *E. coli.* The department says the health risk threat increases considerably as levels rise, especially if water is swallowed accidentally. The guidelines say the presence of *E. coli* can also be an indication of more serious bacterial pathogens such as cholera, shigella, dysentery or typhoid fever (Carnie and Padayachee 2009).

In his book *Brave New South Africa,* Johnson (2009:105) describes the country's water policy as a 'cynical disaster' insofar as 'major outbreaks of cholera [were] directly traceable to failures in sanitation and water supply'. Johnson attributed this to the 'centralising ideology' (sic) of the first post-apartheid water minister, Kader Asmal (1994–1999). In reality, Asmal is best remembered for his decentralisation of national water service responsibilities to ill-equipped and underfunded municipalities, especially in the impoverished former bantustan areas. Asmal also spent the late 1990s regularly rejecting his 1994 lifeline water mandate in the ANC's Reconstruction and Development Programme, because he was advised by the World Bank (Roome 1995) to bring in private sector providers, who would face a major 'disincentive' to invest if compelled to supply a small amount of free water (50–60 litres per person per day) (Bond 2002).

Moreover, and especially pertinent to Johnson's condition was the decentralisation of poor people's sanitation, as Asmal approved a limited-subsidy, co-payment system and hence got very little uptake. Asmal's most important decentralising ideology was evident in his 1994 White Paper, which mandated 100 percent cost recovery from water consumers for operations and maintenance (again following explicit World Bank advice). In its major *Country Assistance Strategy* document, the World Bank (1999) declared that its advice had been 'instrumental' in the 'radical revision' of South African policy. So while Asmal's policies led directly to the rise of *E. coli* in the water system, this was not because of centralising ideology, but rather the kind of neoliberal policies Johnson has long promoted.

In the *New Left Review,* of all places, Johnson (2009:71) argued against alienating foreign financiers, for 'Should the left take measures which frighten those investors away, the result would be an immediate economic meltdown'. And in spite of rhetoric regarding the pioneering roll out of Free Basic Water in Durban, Macleod's strategy was central to the next stage of the cost-recovery approach in South Africa: provide a small token amount for free (based on means-testing of a household's house valuation), but then raise the second block to the point where the overall price soared, resulting in mass water disconnections and serious state-society conflict.

Demand, disconnections, pricing and urinary diversion

It is worth stressing that the basic problem is not personalities or bureaucratic agency, but structural denial of funding. Former water director-general Mike Muller pointed to Durban's failure to invest adequately in water infrastructure expansion in early 2011:

> I really do think in most of our metros, if we don't panic now, if we don't take action now, we will be in a crisis by 2020. We're not going to run out of water, but there are some hotspots. Durban is actually the most vulnerable metro at the moment... I think we need to panic at the right time and the right time is now (Sunday Independent 2011).

Macleod agreed, 'We are at a point where our dams are unable to sustain the current demand over an extended period'.

The leading proposed solution is water rationing, which could begin in 2012 (Savides 2011). Macleod had advocated building a new dam in 2006 at Spring Grove to supply Durban, 'but construction work has not even started', he observed in 2011 (Savides 2011). An additional unspecified dam estimated to cost US$2.5 billion was also proposed, but there was no money to fund it. The main reason for the shortfall was that after early 2006, decisions on Durban municipal investments were subordinated to the construction of the Moses Mabhida Stadium for the soccer World Cup, which subsequently failed to attract sufficient events and crowds to cover its operating and maintenance costs, much less provide the state a return on its US$430 million capital investment (Bond and Cottle 2011).

Nevertheless, had Macleod adopted a different strategy – 'demand-side management' of water – such as luxury consumption taxes on hedonistic consumers and formal restrictions, he might have achieved major savings that would *not* in turn have required new supply-side enhancements. Such measures, however, would have both curtailed water use and dramatically increased the price of water for high-volume residential and commercial/industrial/agricultural users, policies that could certainly lower demand but would also have generated staunch opposition from powerful interest groups.

To date, the main data for critical assessment of Durban's retail water policies and practices come from Alex Loftus's (2004; 2005; 2006) Oxford University doctoral research on household water (including revelations of 1 000 families disconnected from full, high-pressure water supply *every day* in 2003), and a 2005 study of household water price elasticity by Reg Bailey and Chris Buckley. These studies reflected two of the most important demand-side aspects of post-apartheid water: how much it would cost residents to consume, and whether rising prices were offset by sufficient Free Basic Water.

Bailey and Buckley (2005) identified a doubling in the real price of Durban water to consumers who regularly paid their bills from 1997 to 2003. Using house valuations to determine wealth, they assessed the differential impact of the price rises by estimating the price elasticity of water across three bands of approximated wealth (for details see Bond 2010a:456–457). In short, a price elasticity of –0.55 for the third lowest-wealth residents of Durban resulted in a 32 percent decline in per household monthly consumption (from 22 to 15 kl/hh/month), but for the middle- and highest-wealth two-thirds of the city's regular bill-paying water customers, there were much smaller price elasticities (–0.15 and –0.11 respectively). One might reasonably conclude that poor people were, in relation to their ability to pay, overcharged; while wealthier people were undercharged.

In the 6–20 kl/month range of water consumption, the United Nations Development Programme's (UNDP's) 2006 Human Development Report showed that Durban's prices were far higher than other Third World cities (UNDP 2006:91). This fact was of little interest to Macleod. In a 2007 affidavit to the Johannesburg High Court (Macleod 2007) in response to the UNDP's analysis of Durban's rising block tariff, he neglected to acknowledge that the very high relative price increases in the key consumption range (6–20 kl/hh/month) are a problem. Instead, he characterised the increases within the context of a rising block tariff as 'an important part of the legislative framework for acting on the *right to water*' (emphasis added) in spite of Bailey and Buckley's evidence (2005) that the price elasticity was so high that low-income residents had cut back their consumption dramatically (Macleod 2007:11).

Finally, in a third study, 2008 comparative retail water price data were analysed by the Centre for Applied Legal Studies, Centre on Housing Rights and Evictions, and the Norwegian Centre for Human Rights (2008). That study revealed that only Msunduzi (Pietermaritzburg) charged urban South Africans a higher price for a typical low-income consumption level of 12 kl/hh/month at full pressure: R50 (Msunduzi was R55). The price was R35 for a semi-pressure water level. In contrast, Cape Town charges only R10 for 12 kl/hh/month. To its credit, in response to these studies and other factors, Durban shifted its pricing, which extended Free Basic Water. Yet here, too, there was a neoliberal turn, as the new policy incorporated a return to indigence policy, disqualifying people living in houses valued at more than R180 000 (US$28,000) from free services.

South Africa's real estate bubble was so severe during the 1997–2008 global property boom – a 389 percent real price increase (compared to the second highest, Ireland, at half that) – that this simply maintained the exclusion of vast numbers of the poor from desperately needed water and sewerage services. Furthermore, using property value to determine a family's wealth proved wholly inadequate in advancing the desperately needed process of racial and class integration in post-apartheid South Africa (Bond 2010b).

Given the excessive average price increases beginning in the late 1990s, the retail controversies that arose during the early 2000s concerned both the size of the initial block of free water and water disconnections. Why was the Free Basic Water amount of 6 kl/hh/month chosen? In testimony to the Johannesburg High Court, Macleod (2007:3) explained that Durban's 1997 model, which became the basis for the Department of Water Affairs and Forestry's subsequent national policy, was adopted because 'approximately seven litres of water was used per person per day as this was generally the amount that an individual could physically carry and could afford'. Rather than human need, the biophysical limits of water transport – typically by women or children – became the baseline.

But administrative cost factors were apparently also important, and in motivating a free supply of (just) 25 litres per person each day, the national water minister at the time, Ronnie Kasrils, couched his justification in the narrowest of terms: 'It would save money because local authorities would not be saddled with the problem of administering large numbers of small accounts', he told a media briefing in February 2000 (Bond 2002). To achieve administrative savings in billing, Durban therefore began to provide a 200-litre drum at the front door of each shack, which, Macleod (2007:3) said, 'could be filled once a day with clean drinking water … at a minimal charge'. But, Macleod explained (2007:4), during 1998, when 'it became apparent that the amount of money that was collected by the Council for the water supply was in fact equivalent to or less than the costs of administering the collection of the amounts from the relevant communities', Durban began providing the basic amount for free.

Nevertheless, for consumption on a billed meter (not the houses with 200-litre drums), this was an era in which the bulk water supplier to Durban, Umgeni Water, was itself facing financial (and managerial) problems, and its price increases to Durban were passed on to consumers (Loftus 2005). Loftus (2003) questioned Macleod regarding low-income Durban residents' difficulties paying their water bills given these price increases:

> In an interview with Macleod, what angered me above all was his insensitive arrogance towards such situations. He openly admitted to roughly 1 000 disconnections taking place across the municipality daily. This shouldn't be viewed as a problem, he assured me, [since] there is no need to adjust the tariffs or increase the free water allowance. [R]ather the disconnections are (once again) merely a mechanism of financial management. And what about the woman's situation with her "over-restricted flow"? She must have a leak. Now, even I'm enough of a water engineer to know that if she had had a leak, her meter would read a higher level than she was actually receiving. Instead, what Macleod wanted me to understand, was that people lie in townships. They lie about not being able to afford to pay for water and lie about how many people live in their houses. They lie to PhD researchers. These terrible lies have even touched the upper echelons of the city council, the Exco, who Macleod's team are busy convincing of the error of their ways (along with "countering the lies of the enemy – those concerned citizen whatevers").

That 'enemy', the Durban Concerned Citizens Forum, was the initial advocacy group that took Macleod to court to try to halt disconnections (Desai 2002). In 2000, the case of Christina Manqele (a single mother of seven) initially resulted in a court-ordered reconnection after Manqele could not pay her R10 000 water bill and claimed the disconnection was unconstitutional. According to Heinrich Bohmke (2003),

> The Council submitted that just as the state had been unwilling to provide medical treatment to Soobramoney, a refusal which cost him his life as a result of financial constraints, the state could not be expected to supply

water to all who needed it. Wallis [Durban counsel] said the constitutional provisions depended on the availability of resources, placing no obligations on the state and providing no unqualified guarantees.

Pillemer (Manqele counsel) argued that since everyone in Durban was entitled to at least six free kilolitres of water per month, in disconnecting Manqele's water, the Council was denying a person who most needed access to such a free amount. He said Manqele was prepared to use less than six kilolitres a month.

Judge Niles-Duner interjected from the bench to share her memories of the water restrictions during the 1984 drought. She said consumers then were expected to monitor their consumption, otherwise supply was cut off.

In the Durban Metro Council, Wallis argued, there were not sufficient resources to police the use of the six free kilolitres. It was a difficult choice to be made but the court also had to consider 8 000 other households that might want to have their water turned on. For Manqele and her children, it was argued that to deny them access to water because there were no means to ensure that only a basic supply was consumed was but to reinvoke a limitation of rights where the Constitution did not allow limitation in respect of the rights of children.

With more than 10 000 Umlazi township residents in the same situation of having recently been disconnected, the court ultimately ruled against Manqele, finding that by using more than the six kiloletres per family per month then available, she had forfeited any right to free water. However, at that point, Macleod announced that instead of outright disconnections, flow limiters would be installed, an important distinction (Loftus 2005:193). But the limiters were often removed by local water activists (Desai 2002), and for those on the formal metering system, the escalating price of water continued to force low-income people to cut their water consumption, even though cholera and diarrhoea were all rife during this period, and water-borne diseases often took HIV-positive people through to full-blown AIDS.

Further impetus for market-related pricing and disconnections came from the central government. The Department of Water Affairs and Forestry was initially advised by World Bank water expert John Roome (1995:52) to avoid a cross-subsidised water system (especially free water for the first block of consumption) and instead disconnect households that were not paying. The then Minster of Water, Kader Asmal, proceeded to authorise disconnections, including in the former bantustan areas that had received free supplies during apartheid. In 2000, Ngwelezane, a bantustan in rural KwaZulu-Natal, became the epicentre of a major cholera epidemic (Bond 2002) after hundreds of families were forced to draw on polluted surface water supplies. Nevermind that the costs associated with treating epidemic victims were far greater than the benefits in cost-recovery – only around R50 (US$7) per connection from those who could afford to pay.

In addition, the Department of Constitutional Development and the national Treasury established 'Project Viability' in 1997, which encouraged water and electricity disconnections of residents who were in arrears on payments. This project had a perverse outcome: the tougher the credit control, the higher the municipality's credit rating (Bond 2000). Between 2002 and 2003, according to the SA Cities Network (2006), Durban disconnected 40 000 more of its residents than it connected to electricity, and this alongside a strong reserves fund earned the city an AA rating for municipal bonds. The philosophy of disconnection was important, Macleod said in December 2004 (cited in Parikh 2006), because

> We run exactly like a business except that we're not operating as a profit maker or a loss maker; we have to pass on tariffs to recover our costs… The basic reality is that this is not a socialist state, we're not into free gifts, this is no free lunch.

Even without formal privatisation, the *commodification* of water was an explicit objective for the city.

In 2005, the local ECOPEACE party's sole city councillor, Alan Murphy, requested a shift in municipal water policy to write off water-related arrears and redistribute water from industry and high-wealth homeowners to low-income, low-consumption Durban residents. Despite his rhetoric on the need to create jobs and make water available to the poor, Macleod's response (2005) clearly revealed whose interests the city served:

> In terms of the price of water to industrial customers, we are under extreme pressure as there is a generally held view in the marketplace that the price of water to industrial customers in Durban is very high. Research we have conducted shows that, compared to the other Metropolitan areas this is not generally the case, but if one compares our industrial price to an area such as Richards Bay/Empangeni then we are indeed far more expensive than they are. *Any increase in the water price to ICI customers (Industrial, Commercial, Institutional) sends a negative signal to potential investors* and will further reduce our potential to create jobs… We were the first city to introduce the

> concept of free water as a formal policy and have led the way with every step that has been taken in terms of making water services more affordable and more readily available to the poorest citizens in our communities. [Emphasis added.]

Nevertheless, within three years Macleod did expand the Free Basic Water supply to 9 kl/hh/month, with a modified means test that still used property valuation and regrettably dropped the prior universal entitlement to Free Basic Water. The new policy stipulated refilling the 200-litre drums, initially once each day and in future 1.5 times a day. There are various explanations for this laudable policy shift, including the embarrassing findings of Bailey's research; a national 'Water Dialogues' multi-stakeholder process (in which Macleod was a central participant) that more clearly spelt out rationales for increasing water to low-income households (Galvin 2009); and the use of focus groups, which taught Macleod more about consumer grievances (Wilson, Malakoana and Gounden 2008). Bottom-up articulations of grievances, including protests at City Hall over inadequate water and sanitation as well as the rise of localised 'service delivery protests' (Bond 2010b; 2011b) may also have influenced municipal water policy.

Meanwhile, the legal strategies of five Soweto women who fought for a larger water allotment in the Johannesburg courts ultimately failed in September 2009. However, by April 2008 this challenge had generated municipal policy changes that were also soon implemented in Durban. The September 2009 decision of South Africa's Constitutional Court in the case Mazibuko *et al.* v Johannesburg Water *et al.* overturned seminal pro-rights findings in lower courts in April 2008 and March 2009, which human rights activists had hoped would substantially expand water access to poor people.

The Coalition Against Water Privatisation and the University of the Witwatersrand Centre for Applied Legal Studies worked on behalf of women (led by Lindiwe Mazibuko of Phiri) who demanded doubling the amount of free municipal water from an average of 25 to 50 litres per capita per day (50 litres is roughly equivalent to four flushes of a typical toilet) as well as prohibiting pre-paid meter systems, which serve as automatic disconnection mechanisms in low-income black neighbourhoods, in contrast to credit-based meters in higher-income, historically white residential areas. These policies were imposed on Soweto in 2001 by the French company Suez, which until 2006 managed Johannesburg Water, of which Macleod was a board member. The women initially won their case, but the municipality appealed to the highest court where the gains were reversed (Bond 2011a).

In the end, the Constitutional Court refused to become involved in setting policy and implementation guidelines that would give meaning to the following promise made by the ruling African National Congress (ANC) just before the 2000 municipal elections:

> The ANC-led local government will provide all residents with a free basic amount of water, electricity, and other municipal services so as to help the poor. Those who use more than the basic amounts will pay for the extra they use (ANC 2000).

Johannesburg Water officials had in 2001 distorted what could have been a progressive rights-based mandate by adopting a tokenistic free water allocation (6 000 litres per household per month, which for a family of eight is 25 litres per day), refusing to allocate on a per capita basis (hence biasing against large, low-income households), and then raising the second block of consumption to an extremely high level (Bond and Dugard 2008). Activists demanded a different tariff curve, one that would start with a larger lifeline block, followed by higher tariffs rising quickly after a luxury consumption level (typically above 30 kl/hh/month).

In early 2008, anticipating defeat in the High Court (where in April a judge ruled the Johannesburg Water policy 'unconstitutional' and 'discriminatory'), municipal officials changed the policy so that only the small proportion of Johannesburg residents declared 'indigent' would qualify (Naidoo 2010). However, their free water allocation was raised to 10 kl/hh/month. The case is an important precedent in setting out a judicial baseline for water rights, and it thus suggests the limits of what a liberal capitalist democracy in a middle-income society will consider minimal (25 litres/capita/day) and by what means water can be delivered (self-disconnecting meters). In other words, the case very explicitly set boundaries as to what constitutes the 'right to water', even though these continue to be bitterly disputed by communities, such as in Soweto, where residents' illegal reconnections are common (Bond and Dugard 2008; Bond 2011a).

This left Durban's residents, in the wake of the Phiri judgment, with a mix of grievances, which prompted much more high-profile protest activity against other service shortfalls beyond water (Bond 2011b). But the relationship of inadequate services for low-income Durban residents to the inadequate maintenance in the system as a whole, may provide new insights into nature-society linkages, which will have longer-term consequences.

Starting at the level of the household, Macleod (2008a:2) acknowledged that with the city's expansion in 2001, there were 200 000 families – roughly a third of the population – without basic sanitation. Many lived beyond the sanitation belt, a vast peripheral band of the municipality where it was deemed fiscally unrealistic to lay sewage lines. Across the city, despite the fact that there were 60 000 pit latrine toilets that had filled, Macleod (2008a:7) conceded at the 2008 Africa Sanitation conference in Durban that the city was unable to fulfill its commitment to a 'free basic sanitation service in the form of one pit emptying every five years'. Indeed, many pits were unlined with the 'toilets subject to catastrophic collapse', many were 'constructed in inaccessible locations', and there was high variability in content, size, and cost of emptying. The cost of emptying each pit averaged R850 (US$120) per pit compared to 'the cost of constructing new single Ventilated Improved Pit-latrine (VIP) type toilets: R1 000 to R3 000 [US$140 to US$420]', making the process 'uneconomic' (Macleod 2008:7).

By that stage, Simphiwe Nojiyeza and Baruti Amisi (2008:2) noted that Durban still suffered from '148 688 pit latrines without ventilation as well as 41 880 chemical toilets. Bucket latrines have been reduced to 9 270'. As for the VIP system, they

> are designed for Black, poor and rural dwellers whereas full flush toilets are designed for the White and Black Bourgeoisie. Most communities are resisting construction of VIPs which is another form of a bucket latrine (Nojiyeza and Amisi 2008).

VIPs are increasingly filled without being emptied in Durban and across South Africa. As veteran sanitation practitioner Kathy Eales remarked at an AfricaSan conference preparatory meeting in February 2008,

> South Africa's household sanitation policy is grossly inadequate. It speaks primarily to dry systems, and does not clarify roles and responsibilities around what to do when pits are full. National government under-estimated the scale of technical support required (cited in Bond et al 2008).

Interestingly, on a 2010 visit, Bill Gates (2010) was only provided the upside of VIPs:

> Most of us take for granted the convenience and sanitation benefits of flush toilets. But in Durban, many people don't have access to water.

So reducing the incidence of diarrhoeal diseases and worm infections associated with the use of pit toilets is important – especially for young children who are the most vulnerable. Neal showed me the VIP toilet – which isn't as fancy as its name suggests, but is a breakthrough in basic sanitation through the use of simple ventilation methods and other inexpensive construction methods, such as installing a fly screen on the ventilation pipe.

In its *Low Carbon City* assessment of Durban, ASSAf (2011:110) offered a narrow climate-mitigation argument in favour of VIPs and other non-water toilets. The argument against providing flush toilets for poor people was that 'Wastewater treatment plants, which have activated sludge units, have a high impact on [g]reen[h]ouse [g]as emissions due to their energy requirements'. The energy is used for aerobic wastewater treatment, which uses 'large amounts of electricity to drive the activated sludge process' (ASSAf 2011:110).

With VIPs increasingly unattractive to the municipality because of the excessively high cost of emptying them, Macleod (2008a:7) turned to 'the use of double pit, urine diversion (UD) toilets, outside the urban edge'. This in turn permitted further transfer of sanitation maintenance responsibilities away from the state. The simple UD technology entails faeces drying separately, away from urine. But preventing contamination of the former by the latter requires consistent hygienic education. In a context of high rates of diarrhoea (sometimes associated with HIV/AIDS-related opportunistic diseases – a relevant association since Durban has the world's greatest urban HIV-prevalence), humid weather, hilly slopes, poor drainage, and turbulent rain, this contamination is a frequent problem.

Nevertheless, Macleod (2008a:8) maintained that the 63 000 new UD toilets constructed by 2008 were 'able to be emptied by households at an affordable cost' and would use 'minimum amounts of water, if at all'. (Indeed the UD is designed *not* to use any flushing mechanism, and a 2010 innovation was to begin paying residents an estimated R30 (US$4)/month per household for the collection and recycling of urine from UDs as a way to mitigate a 'peak phosphate' problem facing commercial agriculture.) The capital cost of each UD toilet was an average of R3 500 (US$500) the city saves on maintenance costs, because 'emptying is the responsibility of the household, with entrepreneurs already offering their services at R30 (US$4) per chamber emptied'. The city again committed to provide 'extensive community education' and research and explain the 'environmental impacts and health benefits' to enable opportunities for 'local labour and local small businesses [to be] maximised' (Macleod 2008:8–9).

Macleod (2008a:8–11) claimed that 'follow-up visits after construction have increased acceptance levels and emphasised the family's responsibilities for maintenance of the toilet'. He noted the 'period needed for follow-ups extends to years' because of the need to 'evaluate acceptance of the solution and to confirm that the hygiene messages have been internalised'. It was this innovation that most impressed a *Science* journalist (Koenig 2008:744) who termed UD the 'best solution' to Durban's sanitation challenge.

Yet complaints about the UD system were increasingly common beyond the sanitation belt, such as Inanda Dam. As community organiser, Dudu Khumalo remarked about the Umzinyathi and KwaNgcolosi pilot communities,

> UDs have internal buckets and pits that are far too shallow for long-term use. No training was given on how to deal with faeces, except to dump it in the garden 'for fertilising your veggies'. These communities are repelled by human excrement as fertiliser, because of the many diseases surrounding them, compared to cow-dung. The burden of cleaning is left to women. Other creative opportunities for bio-gas are also foreclosed by UDs. Many UDs have become mere storerooms or are permanently locked because of the smells. Councillors are useless when the UDs cease functioning (Bond et al 2008).

Journalists at Durban's *Daily News* picked up the story in mid-2010:

> Some residents of a rural township west of Durban are crying foul over a government project that forces them to dig out waste from their latrines for use as garden fertiliser... The authorities have blamed some residents with smelly toilets, saying that the system was not being used correctly. The toilets are made up of two concrete boxes designed to completely lock out any moisture. Urine is diverted to a chamber separate from that of faecal material, which falls into the concrete chambers.

Once the first chamber is full, the homeowner must seal it, leaving the faeces to decompose while using the adjoining chamber. Once both chambers are full, the decomposed matter is dug out, with the idea that it eventually gets strewn over a garden as fertiliser. Local residents who complained are aghast, not only at the unbearable stench, but the thought of digging out their own waste and using it on their vegetable patches.

'It is not right. It smells terribly and is very near to our house, so we have to live with that smell', said one resident, Nhlanhla Sokhela. Her elderly grandmother, Ntombizodwa Sokhela, who lives with 17 others in her home, all sharing the same toilet, added: 'I do not understand why the bottom is cemented and why we can't just let it sink as with the toilets we built ourselves. How am I to dig out all that waste when it is full?'

Winnie Mkhize, another resident, claimed there had been no training on how to use the toilets and the only equipment left for them was a rake. 'We are 10 living in my house and it takes four months for these toilets to fill up. The smell is not good and it looks very disgusting to have to empty the waste', Mkhize said. 'They say that it is supposed to dry up, but it does not'. The family admitted they had broken down one of the two toilets built for them as the smell next to the house was overbearing.

Particularly irksome, the residents said, was that most of the community had access to piped water, so they felt there should be no problem installing flushing toilets. This was denied, however, by local councillor, Mbongelwa Phetha, who said only part of the area had running water, while others used water tanks (Adriaan, Ngcobo and Mngoma 2010).

There is a substantial difference between the UD device chosen in Durban (without water and with maintenance and cleaning responsibility completely devolved to the household) and UN Habitat's recommended low-cost sanitation system. The latter has various advantages over the UD system, and more than a million have been installed in India:

> The twin-pit system uses 1.5–2 litres of water per use in a flush toilet that is connected to two pits that allows recharging of the soil and composting, and a close-loop public toilet system attached to a bio-gas digester. In fact, this is the only sanitation technology that meets the seven conditions for a sanitary latrine laid down by the World Health Organisation. These stipulate that a sanitary latrine should not contaminate surface soil, ground water or surface water. Excreta should not be accessible to flies or animals. There should be no handling of fresh excreta, or when this is unavoidable, kept to a bare minimum. There should be no odour or unsightliness and the methods used should be simple and inexpensive in construction and operation (Reddy 2007).

As an aside, the matter of sanitation is not only important for residents at home, but for workers or those in transit across the city, especially in commercial areas. In early 2008, there were only 42 public toilets across Durban's entire metropolitan area (not including sports facilities). As journalist Matthew Savides (2008) reported at that time,

> Only four public toilets cater for more than 80 000 people who pass through the Warwick Junction taxi rank, a problem which is replicated at the many taxi ranks across the eThekwini Municipality… The steel structures are shaped like a semi-circle, and the user can be clearly seen while urinating. The urine goes down a drain, but the toilets do not flush, nor do they cater for female users.

Another city official, showed that blame-the-victim was again in fashion:

> Current thinking is to eliminate structures, piping or movable items that can be vandalised, hence the simple "French-style" structures recently constructed at certain taxi ranks. An appeal is made to users of these facilities to do so responsibly, and report vandalism of these facilities (Savides 2008).

In sum, several missing factors made Durban's move to UD sanitation controversial and ultimately unsuccessful:

- Appropriate health and hygiene awareness and behaviour (especially for policy makers and municipal officials prone to disconnect poor people from water supplies);
- systems for disposing of human excreta, household wastewater and refuse, that are acceptable, affordable (to the users), safe, hygienic and easily accessible, while not having an unacceptable impact on the environment; and
- a toilet facility for each household.

To achieve these objectives would require far more than has been allocated in state funding, as well as policies aiming to provide poor people with appropriate supplies of water and water-borne sanitation.

Conclusion

The analysis above is dire in part because expectations have always been high for Durban as a result of professional hype about water and sanitation management. In addition, it was presumed that civic groups would play a strong countervailing role as watchdogs. The hype and awards were, one might conclude, unwarranted. One reason was the failure to budget and spend sufficient funds on low-income Durban residents' water and sanitation needs, as well as on bulk infrastructure expansion and infrastructure maintenance, while another was civil society's failure.

As was becoming clear by the time of the 2010 World Cup soccer tournament, the South African state had vast quantities of funding at its disposal for infrastructure, having spent in excess of US$20 billion on the games, new and revamped stadiums, a new fast train in Johannesburg, and an additional airport in Durban (Bond and Cottle 2011). However, in part because of the attention given to the World Cup and its associated above-ground infrastructure (e.g., in Durban, in addition to the new airport, highways were widened and a flyover was built), municipalities didn't pay enough attention to below-ground infrastructure such as household sanitation and sewage and stormwater pipe rehabilitation, which could have employed many more people and met basic needs. Municipal capital account underspending estimated at US$2.9 billion between 2009 and 2010 'is a worry', the *Financial Mail* (2010) observed, because it implies 'neglecting infrastructure maintenance'. Indeed, 'analysts found there is no budget for repairs and maintenance in most municipalities in 2009/2010. It appears to have been absorbed into the other expenditure line items'. This problem was not only related to the World Cup, as an earlier Treasury (2008) report on municipal capital spending points out:

> Despite the efforts of various capacity-building programmes, municipalities are still underspending national transfers. Some of this may be attributable to the misalignment of national and municipal financial years and the tendency to plan more for the short-term than the long-term. Municipalities tend to report improved expenditure levels towards the end of their financial year. This is an indication of the lack of planning for the execution of their capital spending programmes.

The same problem was also identified by former Director-General of Water, Mike Muller (2007), who addressed the class implications:

> [T]here is an emerging debate about whether decentralisation will achieve the pro-poor benefits claimed for it, since it certainly appears that, with fiscal decentralisation, a greater proportion of transferred funds are being used for "non-basic" services. This often occurs where municipalities have failed to persuade the better-served communities to pay their bills. In South Africa's sanitation sector, while the rate of spending is increasing, the expansion of services to the unserved is slowing, in part due to these effects. The weak conditionality in the Division of Revenue Act, which regulates the municipal transfers, now has clear incentives for municipalities *not to extend services to the unserved.* (Emphasis added.)

In short, we can conclude that South Africa requires a much different national and municipal government policy – one that would incentivise spending on water and sanitation infrastructure so as to prevent the damage being done to the environment and public health, even in Durban, at both the retail and bulk levels. To achieve that will take sustained pressure from civil society.

Looking at the access, pricing, and service-level quality problems, there are two ways to approach the overall failure of low-income civil society organisations to address Durban's water and sanitation management failures. One is to deride urban social movements as ineffectual because they were either *uncivilised* (Wilson 2006) or *too* civilised in their resistance to municipal policy. The most recognised advocate of the latter position, is Ashwin Desai, author of the celebrated *We are the Poors* (Desai 2002). In mid-2010, Desai (2010) backtracked on his own expectations for Durban's urban social movements, as a political force that can right the wrongs of water mismanagement and urban exclusion more generally:

> [H]owever bright the first spark of a usually violent confrontation with the state, the upsurges were siphoned off into calls for participation, legal challenges, and 'voice'. This trajectory saw a decline in upsurges, the eroding of popular support, the emergence of oligarchies and bureaucracy within movements and the need to police one's area in order to ensure the ability to 'speak' for the community…

In order for *de facto* recognition and therefore delivery (one had a constituency to keep happy), in the flat-lands of Chatsworth for example, movement activists had to increasingly join with the city council in various committees. Soon, political work involved technical issues and oversight over upgrading, liaison with welfare departments and a range of other interventions, which pressed less for radical policy change but focused instead on merely getting existing policy implemented. Communities became isolated from each other as the intense specificity of the quest for very local delivery overtook the need for broader solidarities based on broader political principle. In the shacklands, in order for *de facto* recognition from council one had to ensure that no more shacks were built. One also had to ward off competitors. This was especially so if an organisation defined its role as ensuring delivery. Criticism became Janus-faced. Movements took political positions that were severely critical of neoliberal policies on the one hand, while negotiating with them for better delivery within those policy frameworks on the other.

Behind the disappointment over co-optation of urban social movements, however, Desai (2010) also confessed an over-expectation of what was possible:

> The civic as a site of post-apartheid resistance to neo-liberalism has failed dismally to take off. We must consider the reasons for this. Partly it is because of an assumption, seldom made by those actually living in townships, of substantial ground for unity that flows from merely living under the same conditions. The great homogenously marginalised shackland population has failed to congeal as a political force around a common ideology that points to a systemic critique. This is because right now they are not. There, sheer sensuality of the township or ghetto that would make it so ripe for revolution, also makes it so diverse, intersected, fractious and difficult to build a fighting movement once one tries to go beyond the most basic common-denominator demands – such as don't evict us, give us some more toilets.

For most, of course, the toilets did not materialise. Desai (2010) assessed the problem of such narrow 'economism' (criticised by Lenin and Luxemburg), which has been reborn within the post-apartheid urban social movements as 'constitutionalism'. An example was reliance upon lawyers by Soweto residents in Mazibuko *et al.* v Johannesburg Water *et al.*, in which residents suffered a major defeat in 2009 for water policy advocacy (Bond 2011b). Moreover, attempts to 'outmassify' the ruling party would fail, Desai (2010) argued, because '[s]ocial movements won't become mass movements. Let us face it'. They are 'reliant on spin, spectacle and support to exert the kind of influence one needs to be an agent of even minor reforms, not to speak of radical ones'. Moreover, '[s]ocial movements have shown themselves to be as prone to bureaucracy, corruption and leadership for life as any trade union. It is sometimes hard to distinguish movements from NGOs', Desai (2010) concluded.

These are profound critiques, and to some extent they suggest a need, as a result of the disastrous outcome of relying upon a juristic strategy in Soweto (Bond 2011b), to transcend the individualist and consumerist orientation of the 'right to water' narrative, and move to a 'commons' approach. The latter would also permit, as Karen Bakker (2007:436) argues, a tighter integration of the microeconomic consumption battles with the hydro-ecological context in which bulk water is delivered (or not, in times of shortages, underinvestment, and climate change) and sewage carried away and treated (or not, when maintenance is not performed): 'The biophysical properties of resources, together with local governance frameworks, strongly influence the types of neoliberal reforms which are likely to be introduced'. The point here would be for civic activists to campaign more forcefully against luxury water consumption by rich and corporate customers, and by extension dispute Durban's need for expensive bulk-water enhancements.

By linking the environmental and economic in this manner, we may better connect the dots between bulk water collection in the Inanda Dam, the ecological health of Durban's rivers, the various acts of water consumption (and wastage), and the disposal of water into the Indian Ocean. Bakker (2007: 436) worries that 'in failing to exercise sufficient analytical precision in analysing processes of "neoliberalising nature", we are likely to misinterpret the reasons for, and incorrectly characterise the pathway of specific neoliberal reforms'. In Durban we see evidence that the city's finances, including water tariffs, provide the rationale for the various neoliberal policy decisions – but also the potential solution to the multiple water crises if a different regime one day takes power.

Yet notwithstanding the shortcomings of Durban's social movements in addressing these crises during the 2000s, we might argue, nevertheless, that a few real concessions (not reformist cooptation) occurred in the process, including the R35 million (US$5 million) refurbishment and upgrading of infrastructure in Chatsworth in partnership with the Westcliff Flat Residents Association (Bond and Naidoo 2008). Also in mid-2008, the move from 6 to 9 kl/household/month of Free Basic Water was approved. Although this was accompanied by a new 'Indigent Policy' applicable only to households with housing valued under R180 000 (US$25 000) – a means-test, albeit one less subject to stigmatisation – it did lead to a demonstrable improvement in access for hundreds of thousands who rely on the plastic 200-litre tanks of water (hence providing a rationale for more studies such as those of Bailey and Buckley, more Water Dialogues processes, and more stakeholder studies).

But the challenge in coming years, as we have seen, is much more serious, especially as water problems are amplified by the climate crisis. As ASSAf (2011:27) put it, 'Water availability is expected to decrease as a result of increases in temperature and evaporation… [such as] a reduction in annual water flow in the Mgeni catchment that feeds Durban of 157.8 million cubic metres by 2100'. The post-apartheid record is not good when it comes to competition for scarce resources, given the extreme inequality that emerged alongside neoliberal public policies.

In Durban, there is no question that during the late 1990s and 2000s, a dramatic water delivery roll-out occurred, thanks to engineering experience, political pragmatism, and the water department's fiscal resources. Yet we have also seen how this strategy was often self-sabotaged by an austerity mentality pervading decisions ranging from lowest-cost household sanitation to deferred sewage pipe maintenance, as well as the pricing, disconnections and Free Basic Water controversies noted above. A review of the core contradictions associated with Durban's management of bulk and retail water/sanitation suggests that in a neoliberal context amidst the highest rate of income inequality in any large country anywhere, South Africans will have to continue with protests characteristic of uncivilised society in order to achieve more civilised hydropolitical-ecological outcomes.

References

Adriaan, T., W. Ngcobo, and N. Mngoma. 2010. We have to live with that smell. *Daily News*. 25 August.

Academy of Science in South Africa. 2011. *Towards a Low-Carbon City: Focus on Durban,* Pretoria.

Bailey, R. and C. Buckley. 2005. Modelling domestic water tariffs. Presentation to the University of KwaZulu-Natal Centre for Civil Society, Durban. 7 November.

Bakker, K. 2007. The "commons" versus the "commodity": Alter-globalisation, anti-privatisation and the human right to water in the global South. *Antipode* 39 (3): 430–455.

Bohmke, H. 2003. The right to water: The Manqele case. Durban, http://www.africanwater.org/right_to_water__manqele_case.htm.

Bond, P. 2000. *Cities of gold, townships of coal*. Africa World Press, Trenton.

———. 2002. *Unsustainable South Africa*. London: Merlin Press.

———. 2005. *Elite transition*. Pietermaritzburg: University of KwaZulu-Natal Press.

———. 2006. *Talk left walk right*. Pietermaritzburg: University of KwaZulu-Natal Press.

———. 2010a. Water, health and the commodification debate. *Review of Radical Political Economics*

42 (4): 445–464.
———. 2010b. South Africa's bubble meets boiling urban social protest. *Monthly Review*. June: 17–28.
———. 2011a. Durban protests, 2009–10. Unpublished paper.
———. 2011b. South Africa's "rights culture" of water consumption: Breaking out of the liberal box and into the commons? In *Water, Cultural Diversity & Global Environmental Change: Emerging Trends, Sustainable Futures?* B. Johnston, L. Hiwasaki, I. Klaver, and V. Strang, eds. Paris: UNESCO.

Bond, P., B. Amisi, D. Khumalo, and S. Nojiyeza. 2008. The neoliberal loo. *Znet Commentaries*. 19 February.

Bond, and J. Dugard. 2008. The case of Johannesburg water: What really happened at the pre-paid "parish pump". *Law, Democracy and Development* 12 (1): 1–28.

Bond, and O. Naidoo. 2008. 'Partial victories for civil society. *The Mercury*. 13 May.

Carnie, T. 2008a. City refuses to release second probe: Sewage killed fish, says report. *The Mercury*. 18 February.
———. 2008b. City sues over blocked sewer. *The Mercury*. 18 February.
———. 2008c. Skills needed to keep the water running. *The Mercury*. 26 February.
———. 2008d. Durban mops up: Storm knocks out refineries. *The Mercury*. 13 March.
———. 2008e. Durban beaches under threat: Red flag for blue flag beaches. *The Mercury*. 14 March.
———. 2008f. Sewage spills into rivers: Fears over poor water treatment. *The Mercury*. 26 February.
———. 2008g. Council coy as sewage leaks blamed: Bad whiff about fish kill reports. *The Mercury*. 6 March.
———. 2008h. Water courses awash with sewage: Safety threat to city rivers. *The Mercury*. 31 January.
———. 2008i. Durban beaches under threat: Red flag for blue flag beaches. *The Mercury*. 14 March.
———. 2008j. High levels of E. coli found in Durban rivers. *The Mercury*. 31 January.
———. 2008k. Blue Flag body spurns Sutcliffe. *The Mercury*. 27 March.

Carnie, T. and K. Padayachee. 2009. Water quality under microscope: Fears over KZN lagoons and beaches. *The Mercury*. 23 March.

Centre for Applied Legal Studies, Centre on Housing Rights and Evictions, and Norwegian Centre for Human Rights. 2008. *Water services fault lines: An assessment of South Africa's water and sanitation across 15 municipalities*. Johannesburg, Geneva and Oslo.

Desai, A. 2002. *We are the poors*. New York: Monthly Review Press.

De Boer, H. 2007. Embarrassing meaning of "eThekwini" revealed. *Daily News*, 22 August.
———. 2008. New policy proposed for Durban beaches. *Daily News*. 14 March.

eThekwini Municipality. 2011. Beachwater Quality. http://www.durban.gov.za/beachwater, accessed 23 August 2011.

Galvin M. 2009. *Straight talk to strengthen delivery in the water services sector*. Durban: The Water Dialogues, www.waterdialogues.org.

Gates, B. 2010. Simple advances, amazing benefits in Africa. *The Gates Notes*. 18 March, http://www.thegatesnotes.com/Topics/Health/Simple-Advances-Amazing-Benefits-in-Africa.

Johnson, R.W. 2009a. Diary. *London Review of Books* 31 (15) 6 August.
———. 2009b. *South Africa's Brave New World: The Beloved Country Since the End of Apartheid.* Penguin Books South Africa.

———. 2009c. False start in South Africa. *New Left Review* 58: 61–74.

Kasrils, R. 2001. Speech by the Minister of Water Affairs and Forestry at the opening of the Durban water recycling plant. 25 July. Durban.

Koenig, R. 2008. Durban's poor get water services long denied. *Science* 319: 744–748.

Leibbrandt, M. Woolard, I. Finn, A. and Argent, J., 2010. Trends in South African income distribution and poverty since the fall of apartheid. OECD Social, Employment and Migration Working Papers No. 101. Southern Africa Labour and Development Research Unit, School of Economics, University of Cape Town.

Liefferlink, M. and E. van Eeden. 2011. Proactive environmental activism to promote the remediation of mined land and acid mine drainage. Paper presented to the Integrated Water Management conference, Toronto, March.

Loftus, A. 2003. Personal communication, Oxford.

Loftus A. 2004. Free water as commodity: The paradoxes of Durban's water service transformations. In *The age of commodity: Water privatisation in Southern Africa*, D. McDonald and G. Ruiters, eds. London: Earthscan.

Loftus, A. 2005. A political ecology of water struggles in Durban, South Africa. Doctoral thesis in the School of Geography and the Environment, University of Oxford.

Loftus A. 2006. The metabolic processes of capital accumulation in Durban's waterscape. In *In the Nature of Cities*, N. Heynen, M. Kaika, and E. Swyngedouw, eds. London: Routledge.

Macleod, N. 2005. Water arrears write-offs. Reply to Councillor Alan Murphy, Durban, 12 September.

Macleod, N. 2007. Answering affidavit of Neil Macleod, Head: Water and Sanitation of the eThekwini Municipality, 8 January, *Mazibuko & Others v City of Johannesburg & Others*, www.law.wits.ac.za/cals.

Macleod, N. 2008a. Reaching low-income communities: Experiences from eThekwini, Durban. Presentation to the AfriSan conference, February.

Macleod, N. 2008b. We are committed to cleaner rivers. *The Mercury*. 19 February.

Maluleke, S. 2009. Dangers lurking in our sea water: Warm, salty water is an ideal environment for this bacterial infection to spread. *Daily News,* 26 March.

McDonald, D. and Pape, J. 2002. *Cost recovery and the crisis of service delivery in South Africa.* Pretoria: HSRC and London: Zed Books.

McDonald, D. and G. Ruiters, eds. 2004. *The age of commodity: Water privatisation in Southern Africa.* London: Earthscan.

The Mercury. 2011. Durban beachfront revamp at risk from rising sea levels. 6 September.

Muller, M. 2007. Parish pump politics. *Progress in Development Studies* 7 (1): 33–45.

Naidoo. 2010. The Making of "the poor" in post-apartheid South Africa. PhD thesis, School of Development Studies, University of KwaZulu-Natal, October.

Nojiyeza, S. and B. Amisi. 2008. The challenges of eradicating bucket sanitation in SA. Paper presented to the AfriSan conference, Durban, February.

Pullan, G. 2008. Poor quality beaches will be their legacy. *North Coast Courier*. 10 September, http://www.northcoastcourier.co.za/2010/09/08/10-september-4/

Padayachee, K. 2010. Blue Flag rethink for Durban beaches. *Sunday Tribune*. 29 January.

Reddy, A.P. 2007. World toilet summit: Sanitation beyond septic tanks and sewers. *InterPress Service*. 6 November.

Roome, J. 1995. Water pricing and management: World Bank Presentation to the SA Water Conservation Conference. Unpublished paper, South Africa, 2 October.

Ryan, M. 2008. Sewage spill decimates fish. *Sunday Tribune*. 16 March.

Sanpark. 2009. Media release: Crocodile deaths continue in the Olifants River, http://www.sanparks.org/parks/kruger/news/2009/crocodile_deaths.php.

Savides, M. 2008. Taxi rank toilets pushed past the limit. *The Mercury*. 14 January.
———. 2011. Water shortages loom for Durban. *Reuters*. 6 March.

South African Cities Alliance. 2006. *Annual Report*. Cape Town.

South African Department of Water and Environment. 2008. *State of the South African Rivers Report: uMngeni and Neighbouring Rivers and Streams*. Pretoria, http://www.dwa.gov.za/iwqs/rhp/state_of_rivers/state_of_umngeni_02/dams.html.

South African Treasury. 2008. Local Government Budgets and Expenditure Review. Pretoria, http://www.treasury.gov.za/publications/igfr/2008/lg/2008%20LG%20Budgets%20and%20Expenditure%20Review%20%28full%20document%29.pdf.

Sunday Independent. 2011. Water crisis looms. 19 February, http://www.sundayindependent.co.za/water-crisis-looms-1.1029065.

Sutcliffe, M. and N. Macleod. 2008. City Manager's Newsletter: Easter beaches. Durban. 24 March.

Turton, A. 2008. Three strategic water quality challenges that decision-makers need to know about and how the CSIR should respond. Keynote address to Centre for Scientific and Industrial Research, Pretoria, 18 November.

United Nations Development Programme. 2006. *Human Development Report*. New York.

van der Merwe, C. 2008. SA's water market could benefit from more PPPs. *Engineering News*. 10 October.

Water Information Network South Africa. 2009. Public accountability through "Citizens' Voice". Cape Town, http://api.ning.com/files/gLpl-APb6KjU59T9X0XGrDauGUatOIfeUghJVi9iyczKNQ-LrrIem2HjBnu0qcVJ3xnvwZyItcFRPk*yKbUnmg36QuSnWPlY/Lesson20.pdf.

Wilson, Z. 2006. Hydropolitics of water and sanitation delivery: The case of Durban. Presentation to the International Society for Third-Sector Research, Bangkok, 4 July.

Wilson, Z., W. Malakoana, and T. Gounden. 2008. Trusting consumers: Involving communities in municipal water service decision-making in Durban, South Africa. *Water SA* 34, 2.

Endnotes

[1] Urinary Diversion toilets do not use any flushing mechanism and are designed to separate urine from faeces. In the context of South Africa, they typically operate as dry pit toilets and have drawn harsh criticism and resentment from those (poor black and other minority groups) who have been forced to use them.

[2] 'Equitable Share' funding is the central-local subsidy system which keeps municipalities alive. However constitutional interpretations of municipal powers prevent even the powerful Treasury from explicitly directing the grant to support water redistribution in poor municipalities. Notoriously, it is often used for inflated managerial salaries and perks.

Chapter 5

Feeling the Heat in Durban

South Durban Community Environmental Alliance

Durban's climate is already changing. It is a degree hotter than in 1950, and storms are more intense but also less predictable and less frequent, so periodic droughts are more likely. It is also vulnerable to rising sea levels and has already experienced a major storm surge that swept away coastal roads and beachside buildings. While the city is beginning to feel the impacts, it is also stoking climate change and is home to some world-class polluters.

Durban's port is the busiest in Africa and serves as the primary gateway for South Africa's imported crude oil and exported petrochemical products. Big industries are concentrated in south Durban. Next to the port is the Island View bulk chemical storage, which contains an extensive infrastructure of tanks and pipelines – some running inland to Gauteng, while others lead directly, beneath residential streets, to the south Durban refineries, Sapref (jointly owned by Shell and BP) and Engen (controlled by Petronas). Between the refineries is the massive Mondi paper mill. Another 150 smokestack industries, together with toxic dumps and a major sewage works, are located in adjacent areas.

What is now Durban was once a series of interconnected wetlands centred on the bay and stretching from the Umgeni River in the north to the Isipingo River in the south. This was the land of the Thuli clan, who were moved out as the first colonial settlement expanded. With the city centre developing on the north side of the bay, Durban's early industrialists looked to the south for flat land close to the port. The area was occupied by black people, mostly of Indian descent, who were brought to the Colony of Natal as indentured labour to work on the sugar plantations. Once free, they made their living as market gardeners, seine netters and small traders. Being excluded from the city's decision-making, they were easy targets (politically).

The process that dispossessed the once-indentured black people and led to the current industrial landscape, began in the 1930s when the all-white Chamber of Commerce and Industry lobbied the all-white Durban Town Council to zone the entire area for industrial development. They also initiated planning to segregate people along lines of colour. Black people were moved to make way for industrial complexes and crowded into nearby residential areas to create a convenient pool of cheap labour. When the apartheid government came to power in 1948, it modelled industrial planning on the precedents set by Durban, while taking the logic of racist planning to new extremes. In the process, thousands of black people were forcibly removed, restricted, or resettled in south Durban. The result is a patchwork of residential and industrial areas located cheek by jowl. The area is now nicknamed 'cancer valley'. It has a population of close to 300 000 people living in the low-income communities of Clairwood, Jacobs, Isipingo, Merebank, Wentworth, Umlazi, and Umbogintwini. The Bluff – a white and predominantly working-class area – also borders the industrial areas.

These south Durban communities have made the city a centre of resistance to the state and corporate agendas that are leading the world to destruction. They have a long history of resistance to polluting industries. In 1996, they formed the South Durban Community Environmental Alliance (SDCEA) to step-up their campaign and connect with fenceline communities in South Africa's other pollution hotspots. In the 2000s, they responded to calls from local fishermen and market gardeners to support their struggles against dispossession. They also linked with other struggles across Durban – with shack dwellers and street traders, and with campaigns in defence of democratic freedoms.

SDCEA has long understood the struggle against polluting industries as part of the struggle against the drivers of climate change. People are now also feeling the impacts of climate change, and it is the poorest that are the most vulnerable. In 2009, SDCEA hosted hearings on climate change and poverty to hear people's own views on the matter, and this was followed by a series of workshops and meetings. The voices of the people who attended the hearings and the two climate justice workshops held in 2011, are quoted throughout the first part of this article[1] and document the people's analysis of the links between their local struggles and climate change.

The South African government published a green paper on climate policy in late 2010 and intends to publish a white paper before the seventeenth Conference of the

Parties (COP 17) to the United Nations Framework Convention on Climate Change (UNFCCC), which meets in Durban from November 28 to December 9, 2011. The SDCEA, together with its allies in Climate Justice Now! South Africa (CJNSA), is highly critical of these processes and has dubbed the meeting the 'Conference of Polluters'. What follows is a summary of the analysis of those most directly affected by the pollution wrought by the state and corporate sector's industrial policy – those whose views are too often left out of public discourse on climate destabilisation and pollution.

Expanding the port, cramping Clairwood

The port is at the centre of Durban's economy. Over the years its development has led to the destruction of the once-extensive mangrove swamps. Infilling to make land for quays, storage, railways and roads, has reduced the bay to a third of its original size, and the rivers have been canalised and diverted (Scott 2003). Much of the bay is now biologically dead. At the same time, people were forced out of the way of successive expansion projects. The once-expansive area of Clairwood is now boxed in by the port infrastructure on one side and the southern M4 Albert Luthuli highway on the other.

Many families have lived here for 60 years or more and carry a living memory of struggle against this piecemeal dispossession over several generations. Residents recall the significant markers in the process. It started around 1936, when the Durban Town Council put a freeze on residential development. Under apartheid, the urban infrastructure was allowed to decay, but in 1986, the council finally responded to community pressure with the promise of a structured plan to enable coherent residential development.

The people of Clairwood won a significant victory when a sewage system was installed to replace the buckets used for sanitation, but the roads, sidewalks and street lighting were largely neglected. With the transition to democracy, they anticipated a real change in the process of development. But alas, that has not yet materialised. As one community member put it, 'Instead, things went backwards. A new council came to power, but the only difference was the colour of the councillors'.

The lifting of anti-apartheid sanctions coincided with the massive expansion of global trade. This trade is dominated by the big transnational corporations. They now put together products made from components produced in many different places

– wherever is cheapest – and so export goods from one part of the corporate empire to another. To facilitate this global restructuring of production, they demanded quick delivery, and ships were made ever bigger, faster and more energy intensive. By 2007, carbon emissions from ships were estimated at up to 5 percent of total global emissions – which was more than that of the whole of Africa (Vidal 2007).

Durban was swept up in the global trade expansion. Its port capacity was expanded with new wharfs, cranes, and container yards and, most recently, the harbour mouth was widened and deepened to accommodate the latest generation of large ships. Demand for transport from the harbour inland grew explosively. With rail capacity stagnating, the trucking industry expanded rapidly, which added to South Africa's carbon emissions from transport.

Trucking companies use local streets for overnight truck stops and, in the view of local residents, do it to cut costs. In the language of the city's planners, Clairwood is the 'back of the port'. The truckers have invaded the narrow streets and colonised vacant land owned but abandoned by the Durban Town Council. Local people see a process of purposeful neglect:

> *'We are woken at 4 a.m. when the trucks start up. The vibration shakes us awake – you can hear, see, and feel them. Crime is not our greatest fear – being run over by a truck is.'*

They also wash out the trucks, some of which carry toxic substances, and let the effluent water flow into the drains, which lead from Clairwood into the bay.

The people say that this invasion is illegal, but that the law is not enforced. The city says that dealing with the problem is 'a process' but, say residents, nothing happens. This is part of a general pattern of neglect. The city turns a blind eye to illegal waste dumping and to noisy workshops operating in contravention of zoning regulations. In an environment created by neglect, crime thrives.

Local people are still demanding proper planning to restore the residential character of the area. This too, says the city, is a process. Nothing happens. Residents say that 'process' has been turned into distraction. The real agenda is to clear the way for a new round of industrial modernisation through the whole of south Durban: 'If Clairwood goes, Austerville, Wentworth, Merebank, and Isipingo will follow. We need everyone to stand together'.

Making the car culture

Port expansion is also linked with the expansion of the Toyota car plant at the other end of the south Durban valley. Supported by massive subsidies from the Department of Trade and Industry (Flatters 2005), Toyota's Prospecton plant next to Isipingo has been transformed from a local to a global manufacturer producing 220 000 vehicles a year. A large proportion of the components are imported, while over 60 percent of the cars are exported. The port's car-terminal capacity has been increased five-fold since 1994 to meet industry needs, but still imports much more than it exports (Inggs 2009). Cars, of course, burn up the better part of the petrol produced at the refineries. It also takes a lot of energy to manufacture them and yet more to build the roads surfaced with bitumen from the refineries. In the 20th century, the entire urban infrastructure was reorganised for the convenience of the minority of people who owned cars. Public transport largely became run down, making it difficult to get around without a car. Durban sold off its bus service to Remant Alton in 2003 in what was widely regarded as a cheap sweetheart deal. The service deteriorated thereafter, before finally collapsing in 2009.[2] As government itself acknowledges, creating a decent public transport system for all is essential if we are serious about climate change.

On the fenceline

In 1946, the development of the airport in south Durban resulted in most of the surrounding wetland areas being drained. In 1954, Mobil built South Africa's first refinery (now the Engen refinery), and BP and Shell followed with Sapref in 1960. Both refineries were built on land that had been used by local people for market gardening. The Engen refinery is right across the road from people's houses in Wentworth, while Sapref is just a kilometre away, across the Umlazi canal from Merebank.

The Key Points Act, a notorious piece of apartheid security legislation, made it illegal to publish anything about the refineries. Not even the local authorities were allowed to know how much pollution the plants were pumping out. Permits issued by the Department of Environment, which regulated only for sulphur dioxide emissions, were similarly secret. In fact, the permits were mostly written by the industries themselves. Engen was allowed to emit 72 metric tons of sulphur dioxide a day, while Sapref was allowed 50. Both claimed to operate well below their permit levels, but in 2000, it emerged that Sapref had cooked its figures. It was, in fact, exceeding its very generous emissions permit allowance (O'Connor and Hallowes 2002).

Under sustained pressure from SDCEA, both plants have substantially reduced their sulphur emissions. It seems, however, that there's a trick to it: They put the surplus sulphur into the heavy diesel used by ships, so it is emitted at sea (Hallowes 2011). Emissions of other toxins, particularly benzene, remain extremely high (Naidoo, Gqaleni, Batterman and Robins 2006). This is from normal operations.

Fires, explosions, gas leaks, spills, and excessive flaring occur with appalling regularity at the petrochemical plants. In 2001, Sapref spilled between 1 million and 2 million litres of fuel from a pipeline buried under a residential street (O'Connor and Hallowes 2002). The spill forced the evacuation of local people and marked the beginning of a lengthy struggle to make Sapref replace its 40-year-old pipes rather than simply patching them. This was just one of 26 spills from the Engen and Sapref refineries recorded by SDCEA from 2001 to the end of 2004.

2007 was a year of fire. At the Island View chemical storage on Durban docks, a series of explosions ripped through eight tanks, which burned through the night of September 18. The air was thick with chemical smoke and fish turned up dead in the water a few days later. More fires at the refineries – three at Engen and one at Sapref – spread fumes and soot across the neighbourhood. In 2008 another major fire at Engen's crude-oil feed shut the entire refinery down for weeks.

The city says it has adequate emergency plans. SDCEA has repeatedly asked to see them but has been refused. People do not believe they really exist:

> *'Incidents and accidents – the fires – they call them "acts of God" but they are not. There is no emergency plan. People don't know where to go when there's a fire. And the old people can't go anywhere without assistance. So we want an emergency plan.'*

People's immediate concern is health. Government and industry have always claimed superior scientific knowledge to belittle this concern. The first systematic study of health in south Durban confirmed that the people were right. Living here is likely to blight your life. Respiratory diseases like asthma are well above the average rate, and people who live with pollution do *not* get used to it – they are made more vulnerable to it. In addition, the chances of getting cancer are 250 times higher than normal.

Most families have lost someone:

> *'I live with the pollution from Engen. I gave birth to a boy in 1993. After fourteen years, he was diagnosed with leukemia. He died on January 13, 2009.'*

Leukemia is strongly associated with benzene (groundWork 2003).

Expanding

Carbon dioxide emissions from the refinery rise or fall pretty much in line with production. One sure way to reduce them is to cut production. Since 1990, both refineries have dramatically expanded production capacity – Engen from 67 000 to 150 000 barrels a day and Sapref from 120 000 to 180 000 barrels a day. The refineries also supply the basic feedstock to south Durban's other chemical industries, many of which are located at the AECI complex in Umbogintwini.

The latest expansion is driven by the government's infrastructure development programme. The state-owned transport and infrastructure corporation, Transnet, is building a high-capacity multi-product pipeline to carry crude oil to Sasol's Natref refinery, and petrol and diesel from the Durban refineries to the greater Gauteng market (Department of Public Enterprises 2007). It is routed through poor semi-rural areas where little resistance was anticipated. But people are angry:

> *'They are taking the pipeline through our gardens. What will happen when there are leaks and explosions? We are sick already. Why are they taking it through our area? They don't talk to us. They don't care about us, because we are poor.'*

In 2010 the new King Shaka International airport was opened, leaving the old airport in south Durban vacant. This was long anticipated, and in 1998 a Strategic Environmental Assessment (SEA) looked at what to do with the land. The Department of Trade and Industry pushed for a 'world class' chemicals cluster, supposedly free of pollution, to take advantage of feedstock from the refineries next door and create jobs. SDCEA did not buy this: Promises of clean production are broken everywhere. Moreover, new investment in Durban's chemicals industries has replaced labour with automation in their plants, which has destroyed jobs rather than create any jobs.

Transnet lobbied for a dugout port to expand capacity even further. This new port would be physically separate from the existing port, and would be located at the opposite end of the valley. For SDCEA, it became evident that this would create a new 'back of the port' hunger for land. The SEA report indicated that port infrastructure would elbow into Merebank just as it does into Clairwood. And the new infrastructure would be required to connect the new port with the old and therefore run straight through all the communities in between. The new port would also destroy what remains of the Isipingo lagoon and mangroves.

In 1999, thousands of people packed into public meetings to denounce these plans. The government, however, expanded its ambitions. It wants the port, the chemicals cluster, plus an 'automotive supplier park' to supplement Toyota next door (Naidoo 2011).

Regulation of industry, meanwhile, has been gutted. Under pressure from SDCEA, Durban's City Health unit painstakingly built up a local air-quality monitoring and regulation system – not perfect, but probably the best in the country. However, in 2011, the number of staff was cut by more than half, effectively restoring industry's charter to pollute.

A growing stream of toxic waste flows from the expanded production. Two toxic dumps were opened in black residential areas in south Durban in the 1980s. Waste-Tech's Umlazi dump did not even have a lining. It leached poisons into the Isipingo River below, while toxic fumes from the dump repeatedly forced nearby schools to shut down. Sustained community protest forced the closure of the dump in 1997, although it continued to take toxic ash from Mondi until 2007 (groundWork Report 2008).

The Durban company Wasteman's Bulbul Drive dump in Chatsworth opened over the protests of the local community. Poorly constructed on steep land, the dump subsequently 'slipped, releasing toxic clouds into the air and a toxic flood into the Umlazi River' (groundWork Report 2008). As at the Umlazi dump, children at local schools are periodically overcome by fumes from the dump even when it is operating 'normally'. Meanwhile,

> *'the waste trucks pass our homes and shops and schools endlessly every day. What they bring in includes galley waste from ships, waste from Rainbow Chicken's industrial farming operations, and high-hazard chromium waste.'*

In 2009, Wasteman applied to expand the dump and extend its life to 2021. Building on the long-running campaign to close the dump, local organisations united to oppose the application. They won. In March 2011, it was finally announced that the dump would be closed by the end of the year, a process that will be closely monitored.

World-class appropriation

Durban's street vendors observe that:

> *'COP 17 will be a big event with many visitors from around the world. And we know that they will then start with "street cleaning," so the international visitors will not see dirty street vendors. We want support for our demand that we are not cleaned off the streets. We have learnt that the same thing happens in other parts of the world when they host big events. And we experienced it during the football World Cup.'*

South Africa pulled off the 2010 World Cup with style. The grand stadiums were all completed on time – though not within budget – and Durban opened its new King Shaka International Airport shortly before the opening. Government said this was proof that it could deliver. The poor of Durban asked, 'To whom?'

In the late 1990s, Durban managed an innovative and generally supportive approach to street traders. This was reversed when the Durban City Council adopted the 'world-class city' slogan, meaning that high-value locations should look attractive to investors. Global capital has made the succession of mega-events into significant sources of profit with carbon emissions to match. For host countries and cities, they present the opportunity to market themselves to global investors. These opportunities come at a very dear cost. FIFA[3] walked off with staggering profits from the World Cup, the corporate sponsors latched onto the global audience, and Durban was stuck with a R2 billion bill. The prestigious Moses Mabhida Stadium has not been filled since the World Cup, and maintenance alone drains the city coffers. South Africa will also pay handsomely for the privilege of hosting COP 17.

The whole of Warwick Triangle, a major commuter terminal and the centre of Durban's street trade, was to be 'cleaned up' ahead of the World Cup. The Early Morning Market, Durban's original fresh produce market, was about to celebrate its centenary year. Instead of celebration, the city targeted it for demolition. It planned to use central government funding for World Cup-related infrastructure to replace the venerable building with a shopping mall that was intended to capture the commuter market. The plan threatened around 2 000 people – stall holders, workers, and barrow boys – all of whom struggle for meagre livelihoods: 'This is the politics of the council, robbing poor people of their livelihoods, destroying small businesses to give it to big business.'

The market was the centre of a web of connections across the city. One community member describes its importance:

> *'The whole of Springfield was farmed when I was growing up, and all the produce went to the Early Morning Market. My grandmother started a stall and handed it on to my parents. That is what kept me and six siblings.'*

Springfield is on the banks of the Umgeni River and has since been turned over to industry and stores selling cheap, imported consumer goods. The farmers were moved and moved again. The last of them are on the old airport land. They still supply the local markets but have been working under the threat of eviction, with their leases extended one month at a time, since 2005. One local citizen describes the historical context and impact of the situation on a close family member:

> *'My grandfather was an indentured labourer. He started farming at Springfield Flats. That was very fertile land. Durban City Council said they needed the land for housing, but then they built industry there. He was moved first out to Phoenix then to the new airport land. Now where?'*

Aside from the poisoning and expropriation of their land, farmers have another worry. Already they feel the climate changing:

> *'The weather is odd. The rain does not come when it is expected. The droughts dry out the land and then the floods come and wash it away. We also have more pests.'*

Because climate change threatens food production, the farmers say it is vital to keep productive land for agriculture close to the city. The municipality, however, is turning it all over to industry. This will undermine the resilience to climate change of the entire community of Durban. 'If we are moved, you'll be paying exorbitant prices for your vegetables at supermarkets', one farmer notes. Out along the route of the new pipeline, small farmers want government to support them in adapting to climate change. What they experience is a government that cannot distinguish support from command: 'Government should give us the seeds we need – not the ones they think we need'.

The street traders are also threatened:

> *'We are concerned that the changing dynamics of the market are squeezing us out. Seventy percent of street trade is in foodstuff but, for example, it is increasingly difficult to get bananas, because they are now bought in bulk. We are trying to respond to that, but we get no support from government, NGOs, or civil society.'*

The city colludes with capitalist market forces using the police 'day-to-day to shut us down' and providing neither shelters nor toilets. And women traders have to look after children while working, because the city will not provide crèches.

The fight to defend the Early Morning Market drew support from community groups across the city. They took to the streets and took the city council to court four times. So far they have stopped the city from closing the market, though, one supporter notes, at great cost: 'We won in the courts but it is very expensive, and we have problems getting legal support'. The council abandoned the project when it became clear that it could not push it through in time to capitalise on World Cup funding. It no doubt awaits its next opportunity.

Subsistence fishermen are also criminalised:

> *'We've been thrown out of the bay and surrounding areas that were our traditional grounds. The harbour is being privatised for industry and polluters. Rules are imposed on us limiting our catch ... They blame shore fishers with one hook for the decline in stocks. They don't look at the big industrial fishers with long lines and kilometres of net. They are the ones depleting the stock'.*

Stocks are also declining with the loss of habitat. The bay and the local estuaries – vital spawning ground for several marine fish species – have shrunk, and most of the mangroves are gone. They are also poisoned by pollution:

> *'Engen uses storms as a cover for releasing polluted waste water. It also pollutes the mouth of the Umlaas canal. That used to be good fishing, but there is nothing there now'.*

The fish themselves are also polluted, so the poison 'comes back in food' to add to the toxic burden carried in people's bodies.

To bring the World Cup to the people and leave a legacy, several township football grounds were upgraded to FIFA training standards. People living next to the King Zwelithini Stadium were moved, one community member offers, 'because they said the place must be clean for 2010 when the visitors come'. The visitors did not come. Some of the people lost substantial homes and were compensated with a one-room house in a toxic area:

> *'We have been removed to where chemicals were dumped. The chemicals are now coming back up. The wind comes into our homes, and our children are sick'.*

Shack dwellers

New people are arriving every day in South Africa's crowded shack settlements. They come from the farms, from the destitute former bantustans (the homelands created for black South Africans during the apartheid era) and from other areas of the city hoping for work. Abahlali baseMjondolo, the shack dweller's movement that started in Durban, has resisted the world-class city agenda to remove them to the peripheries. It sees the roots of dispossession in the unending privatisation of land initiated under colonial rule:

> *'In places where we had houses, they are taken from us, because we do not have the money to bribe officials. And we have no access to RDP [Reconstruction and Development Programme] houses for the same reason'.*

In the towns and cities, they are occupying the spaces left open: 'That is the land that is not privatised, that we have taken back for the people, because we understand that land is a gift from God'. This is often land without value to the market. In the small town of Ixopo, inland from Durban, people have built their shacks from mud. They call it Chocolate City on account of its physical appearance. The lack of infrastructure and services in Chocolate City make it a dangerous place:

> *'It is on a steep slope. There are no proper drains, so it is badly eroded, and we have mudslides when it rains. Whole houses are swept away and people die. We recently buried two children'.*

Shacks are more commonly made of tin, which has its own problems: 'In the heat of summer, people die in their homes'. The shacks are cold in winter and sodden when it rains.

Most are not serviced:

> *'We have no electricity so we must use dirty and dangerous energy to cook, to keep warm, to have light. We have the mbaula [braziers], which fill our homes with smoke, and we use paraffin. We live in constant danger of fires'.*

Respiratory infections from indoor pollution are a leading cause of death in young children. The fires are alarmingly frequent and spread rapidly through the densely-packed shack settlements. Many people lose their lives every year, and many more lose all their possessions.

The people see safe connection to electricity as the most significant solution. However, millions are denied this basic service, and even for those who are hooked up to the national electricity supplier, Eskom, (that is, around four million people), nearly a third have zero consumption recorded because they cannot afford to buy the power legally:

> *'Electricity is for some but not for us, so many of us connect illegally. This, too, is dangerous and people die. And they make us into criminals'.*

Disconnections by utility officials are frequently accompanied by violence: 'The poor in Durban have been abandoned to fire, left to burn, because we do not count in this city'.

Most shack settlements have no water and no sanitation. One shack dweller describes what this means:

> *'This is the most disgusting problem. People defecate in open ground and in rivers. Some use plastic bags which they throw away. People also keep animals to survive and have pigs and chickens living in their homes. Animal and human faeces are mixed. So we are put in the way of disease, and we contaminate the rivers which people downstream use for drinking'.*

Waste pickers

Waste pickers work in dirty and dangerous conditions, mostly for little return. They are among the most vulnerable of people working in the informal economy. But they are getting organised:

'We've founded co-ops and started provincial movements and a national movement. These are our own organisations – of waste pickers, by waste pickers and for waste pickers. Our first concern is getting recognition, because neither government nor communities recognise us or the value of the work we do'.

The dumps produce landfill gas (LFG) – dirty methane (a major greenhouse gas), in particular. This is created from rotting organic matter, and it is dirty because it is contaminated by everything else that goes to the dump:

'The waste management system neglects separation of waste, so it is all mixed together and creates dirty and dangerous working conditions for us. We also appeal to people to separate wet compostable waste from other waste'.

At COP 16 in Cancún, South Africans joined the Global Alliance of Waste Pickers to campaign against false climate solutions and demand recognition for their role in mitigation. Recycling produces the largest savings on carbon emissions: 'We return materials to industry and the avoided emissions along the production chain are even greater than the emissions from dumps and incinerators'.

Waste, like everything else, is also being privatised: 'Waste pickers are displaced so that private companies can take the profits from recycling. Corporations are also promoting incinerators and other forms of waste-to-energy [technology]'. Incinerators are also being promoted by governments and the World Bank as a means of saving on carbon emissions. But as one waste picker points out, this is a false solution to the climate problem: 'It wastes resources that could be recycled'.

At Durban's Bisaser Road dump, dirty LFG is used to produce electricity. But it also produces toxic emissions. It is nevertheless credited under the Clean Development Mechanism of the Kyoto Protocol, because of its claim to save on carbon emissions when compared with the equivalent amount of coal-fired power. New coal-fired stations are built anyway, while the supposed emission reductions are traded away to polluting corporations in the developed countries. These corporations can then claim the reductions without actually reducing their real emissions.

Shack dwellers get no waste services:

'We know about separation but no-one comes to collect. So when we separate we end up throwing it in the dump or even burning it'.

For their part, the waste pickers 'are promoting residential recycling centres where residents can take separated waste'. Beyond this, they are looking for solidarity with community organisations and the movements of the poor, as they come from the same communities.

Exclusion

'We are crying and no one can hear. Nothing has changed. We are the underclass.' This is the common cry of the people that government disregards. The government promises people's participation but it acts in the interest of corporate capital. One citizen describes it this way:

> *'Government talks about women on women's day. They talk about fighting poverty. They say Batho Pele [people first]. But they should change that. For government, it is Money First'.*

People from across Durban try to engage government but meet obstacles at every turn:

> *'We have approached the authorities – both national and local. But if the request is from the shack dwellers, they send people who have no authority. Then when we make a press statement, they will say we were not aware of this problem'.*

Even the formal mechanisms of participation are used to prevent real participation. As in Clairwood, people spend time engaging in processes that never end and never produce any results. Other processes, such as the environmental impact assessment for the multi-product pipeline, merely rubber-stamp the plans on which decisions have already been made.

Shack dwellers in Kennedy Road spent years trying to work within the proper channels of participation, but they were never heard.

> *'So we have taken our protests onto the streets. But we are met with violence. They use rubber bullets and tear gas against us. The system that is supposed to protect us kills us'.*

This is the experience that led them to form Abahlali baseMjondolo.

Protest is itself a legitimate means of participation. In terms of the law, people have to notify the police that they intend holding a protest. They do not have to ask permission, but the police often respond as if they do. So police denied permission when SDCEA and 20 other organisations organised a protest to stop the pipeline. Since the police acted in breach of their own regulations, the people protested anyway.

People in Durban do use the courts to defend their rights and sometimes they win. It was Abahlali who established in court that the police did not have authority to deny them permission to march. Nevertheless, 'the judicial system is biased against the poor'. This is not only because the cost of going to court makes it prohibitive without financial assistance, but also because officers of the court tend to share the prejudice of the establishment against the poor. A man in a suit is generally given more credibility than a woman without shoes. Beyond this, the privatisation of land and labour is backed by the laws of property made by, and for, the rich.

People's demand for participation is not only about whether they are consulted in this or that process about precinct plans or service delivery, or even national policies. Their slogans are, 'Nothing about us without us' and 'Talk to us, not for us'. Real participation means that development as a whole should be democratised. This raises the challenge of beginning to think about what people's development would look like: 'How would we organise production and consumption; what are people's technologies as opposed to corporate technologies; and how would we relate to the environment?'

Where SDCEA stands on COP 17

SDCEA expects little from the governments gathered at the Durban COP. They have not shown themselves to be serious about addressing climate change. They have not shown themselves to be capable of leading the necessary change in the economic system of production and consumption.

SDCEA expects much from the people. On the streets of Copenhagen and Cancún, people chanted 'change the system not the climate'. This is the real issue. This is what will be debated in the people's autonomous space in Durban. This is what people from south Durban, together with their allies from around the city and the world, will call for as they join the Global Day of Action on December 3, 2011.

Changing the system is necessary because capitalism is not compatible with addressing climate change. Capitalism requires never-ending economic growth, which can be sustained only with growing carbon emissions. Addressing climate change requires a very steep reduction in emissions starting *now*. Further delay means steeper reductions.

Governments North and South have adapted their concept of development to Capital's requirement for economic growth. They claim that economic growth is necessary to 'pull people out of poverty'. Growth has indeed brought incredible wealth to the owners of capital and prosperity to the world's middle classes. But it has also brought untold misery to the majority of people, particularly in the global South.

Economic growth is accompanied by the growing inequality of incomes globally in most countries. It also brings intensified pollution and carbon emissions along with large-scale dispossession of those who stand in the way of 'development'.

In South Africa, the richest 20 percent of people take 75 percent of all income, and 60 percent of the people are poor by any reasonable measure. This is not for want of economic growth. The boom years up until 2008 merely intensified poverty as ever-increasing food prices cut into the budgets of poor households.

Targets

There is no *safe* level for rising temperatures or carbon concentrations. The global temperature increase is now about 0.85 degrees Celsius above pre-industrial levels – already catastrophic for millions of people around the world. In 2010, Pakistan drowned while Russia burned. 2011 opened with unprecedented flooding in Australia and Brazil.

In much of Africa the temperature rises at one and a half times the average rate of increase global and already exceeds 1 degrees Celsius. In Niger, several years of drought were followed by unusually severe flooding in August 2010. People – already vulnerable to malnutrition – saw their crops destroyed and their homes flooded. In South Africa successive years of heavy flooding in the southern Cape yielded to severe drought in 2010, while the normally dry Northern Cape Province was inundated with flood waters in early 2011.

The Cancún agreement sets a target for stabilising temperatures at 2 degrees Celsius. This is a target for global catastrophe. The risk of runaway climate change – the point when natural feedbacks become more significant than man-made emissions – is already evident and becomes a near certainty at 2 degrees Celsius.

It is therefore imperative to keep warming as little above one degree as is now physically possible. That probably means 1.5 degrees Celsius which is the target demanded by African countries. This is not a 'safe' target. It is what the global elites have brought us to.

The rising temperature is dragged up by rising concentrations of CO_2 in the atmosphere. As with temperature, 'safe' concentrations are far exceeded. Over the last million years or so, CO_2 concentrations have fluctuated between 180 and 280 parts per million (ppm) in the atmosphere. The concentration is now over 390 ppm, well outside earth's normal operating range, and increasing at around 2 ppm a year. [4]

Including all greenhouse gases, concentrations are now around 440 CO_2e (carbon dioxide equivalent) ppm. Global leaders talk of 'stabilisation' at 450 CO_2e ppm. This does not correlate even with the dangerous 2 degrees Celsius target but puts us on the path to 3 degrees Celsius. Nor have global leaders taken any credible action even to achieve stabilisation at 450 ppm (Anderson and Bows 2011).

It is now clear that the rise in temperature cannot be reversed. If we go to 1.5 degrees Celsius, that is what we are stuck with. But the concentration of CO_2 can and must be reduced to stop the temperature rising further. The 2010 World People's Conference on Climate Change meeting in Cochabamba demanded a return to pre-industrial concentrations: There is no 'carbon space' left. Meeting any credible target now requires that global energy and industrial emissions are cut by between 2 and 3 billion metric tons each year. The implications are evident:

- All oil, coal, and gas exploration should stop – more than enough to burn the planet has already been found;
- extraction of unconventional fossil fuels (tar sands, deep-water deposits, shale gas, coal-to-liquids, etc.) must be abandoned; and
- no more coal-fired power stations should be built in South Africa or anywhere else in the world.

The South African government is not planning for the real future. The National Climate Change Response Green Paper warns that average temperatures will be 3–4 degrees Celsius hotter at the coast and 6–7 degrees Celsius hotter inland. That is not liveable.

But government cannot face up to what it sees coming because it remains wedded to the dominant interests of the minerals-energy complex. It remains locked in a view of the world in which economic growth constitutes the central organising principle of development. This is not because growth is needed to alleviate poverty, but because it is needed to reproduce capital. This is the system that puts profit before people.

To address climate change and meet the needs of the people, there must be a radical redefinition of what is meant by development and who defines it.

1. The central organising principle should be sustainable development founded on economic, social, and environmental justice. This means a commitment to growing human solidarity and equality and that people recognise themselves as a living part of the earth's ecology. To destroy the environment is, finally, to destroy the people.
2. Localisation is essential to any serious programme of mitigation, and requires that national resources are focused on supporting people's capacities to direct local development.
3. The energy system must be transformed as a matter of urgency. This is not only about choosing renewable technologies. It is about what energy is for and who controls it. We call for people's energy sovereignty founded on democratic and local control.
4. The transition to a different energy and development order will require energy inputs from the declining fossil-fuel system. If these investments go into the declining system, they will represent a permanent loss. In the period between now and the latest credible peak-emission target date of 2015, fossil fuel resources should be used to build the new system.
5. Food is the most basic form of energy for people, and the food system must be thoroughly transformed to enable people to define and take control of production and consumption, and hence of their own futures.
6. Finally, the Green Paper repeats government's stated commitment to people-centred development. That commitment is not visible to ordinary people. We believe that a people-centred approach means an open-ended process of transition to a society in which people are actively and consciously making the decisions that shape their collective future.

References

Anderson, K and A. Bows. 2011. Beyond dangerous climate change: emission scenarios for a new world, *Philosophical Transactions of the Royal Society* A, 369, 20–44.

Bulbul Drive Action Committee (statement). 2011. Bulbul Drive dumpsite closed. *Look Local*, August 10, awww.looklocal.co.za

Department of Public Enterprises. 2007. *Strategically Important Developments*, May.

Flatters, F. 2005, The Economics of MIDP and the South African Motor Industry, Queen's University, Canada, prepared for TIPS/NEDLAC South Africa Trade and Poverty Programme (SATPP) Policy Dialogue Workshop, Johannesburg, 2 November.

groundWork. 2003. National report on community based monitoring in South Africa: Air pollution in selected industrial areas 2000–2003, www.groundWork.org.za.

———. 2008. The groundWork Report 2008: Wasting the Nation: making trash of people and places, written by David Hallowes and Victor Munnik. 43.

Hallowes, D. 2011. *Toxic Futures: South Africa in the crises of energy, environment and capital*, UKZN Press, p189.

Inggs, M. 2006. Durban port expansion continues, despite big drop in demand, *Engineering News*, June 5. See also Peter Hall and Glen Robbins, 2006, *Which link, in which chain? Inserting Durban into global automotive supply chains*, UKZN School of Development Studies, Working Paper No 46.

Naidoo, R., N. Gqaleni, S. Batterman and T. Robins. 2006. South Durban Health Study, Centre for Occupational Health, University of KwaZulu-Natal; Department of Environmental Health Sciences, University of Michigan; Department of Environmental Health Sciences, Durban Institute of Technology.

Naidoo, S. 2011. Bigger plans for old airport, *Independent Online*, February 23.

O'Connor, R and D. Hallowes. 2002. *Ground-zero in the carbon economy: People on the petrochemical fenceline*, in Mark Butler and David Hallowes (eds.), *South African people and environments in the world market*, WSSD booklet series, groundWork. 11 and 12.

Scott, D. 2003. Creative Destruction. Early Modernist Planning in the South Durban Industrial Zone, South Africa. *Journal of Southern African Studies* Vol. 29, No. 1. March, 235–259.

Vidal, J. 2007. CO_2 emissions from shipping twice as much as from airlines. *Guardian*, London, March 3.

Endnotes

1 None of these quotes are attributed to the individual speaker, partly because they are from group report backs or are anonymous comments from the floor at the climate justice workshops and hearings. They are meant here to provide a collective analysis.

2 See, for example, Helen Zille zaps ANC 'cronies', Independent on Line, February 4, 2009; IFP KZN Premier Candidate's Press Conference on Transport Statement, March 24, 2009; Durban suspends bus service, Mail and Guardian, May 14, 2009; Statement from eThekwini Mayor Obed Mlaba: Notice from Remant Alton – suspension of bus commuter service, March 13, 2009.

3 FIFA, the *Fédération Internationale de Football Association*, is the parent organisation of the World Cup.

4 See www.co2now.org; or go direct to the US National Oceanic and Atmospheric Administration at www.esrl.noaa.gov/gmd/ccgg/trends/.

Durban's sports tourism precinct
Nkosi Albert Luthuli
International Convention Centre
Hotels
Surfing piers
Suncoast Casino
Two stadiums
The ABSA Stadium Durban
ABSA

Central Durban from Warwick Junction and DUT (top left) to Moses Mabhida Stadium and ICC
Photo: Image Africa

Des D'Sa at Settlers Primary School community meeting
after the Engen oil refinery fire in October 2011
Photo: Patrick Bond

Indian Quarter Street scen
Photo: Patrick Bond

SA police station and Venture Africa building
Photo: Patrick Bond

Bisasar Road landfill, Kennedy Road informal settlement and Umgeni Road
Photo: Jason Hickle

Nkosi Albert Luthuli International Convention Centre & Hilton hotel
Photo: Patrick Bond

Mike Sutcliffe
Photo: Avusa Media

Durban Social Forum protest against the FIFA World Cup
Photo: Vanessa Black

Protest by fishermen against eviction from central piers
Photo: Vanessa Black

Bisasar Road Clean Development Mechanism electricity generators
Photo: Patrick Bond

Grey Street Mosque
Photo: Patrick Bond

Map of South Durban Basin

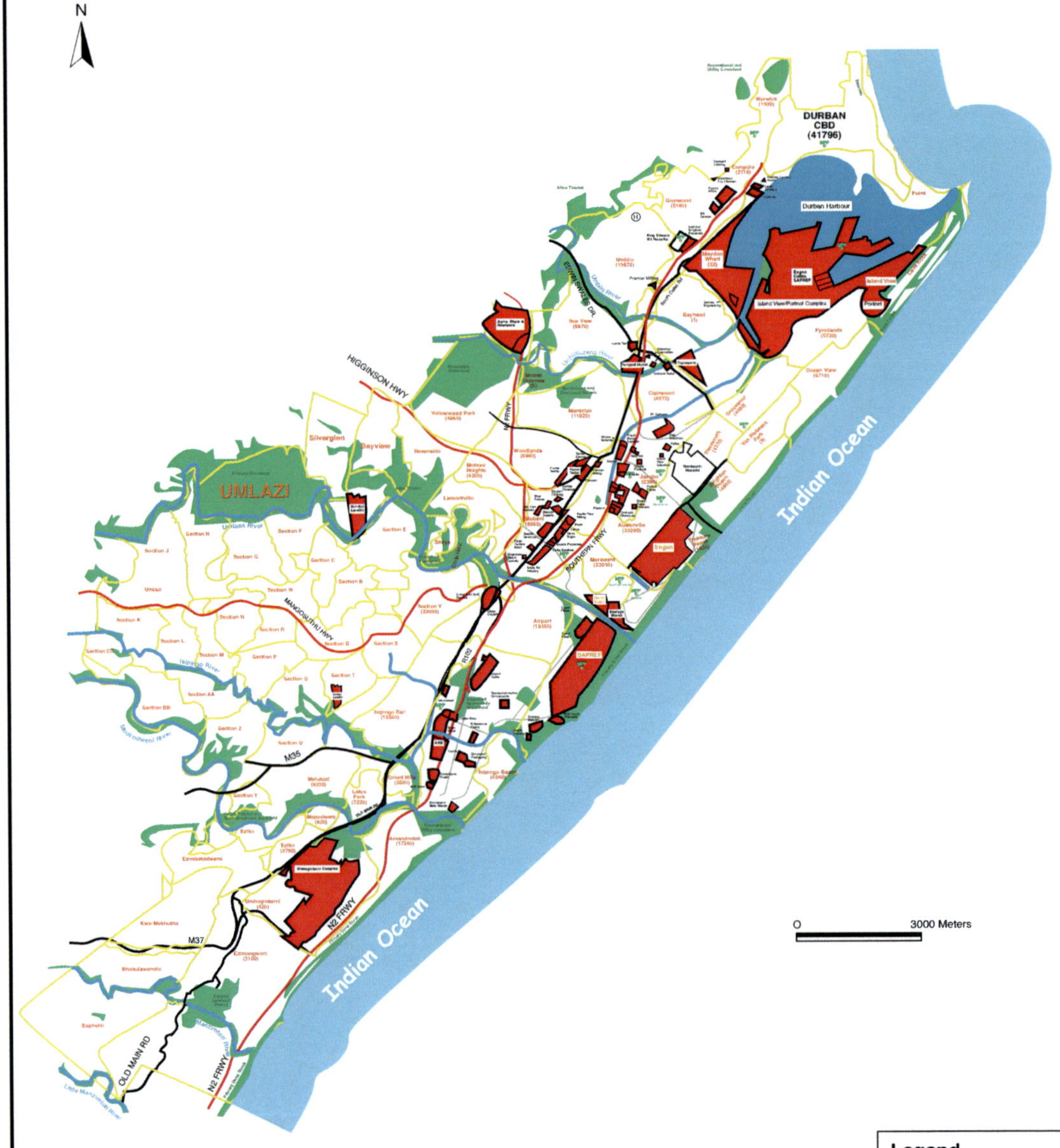

Funded by DANIDA (Danish Ministry Of Foreign Affairs) as part of the SDCEA Local Action Project

SDCEA

The contents of this publication may be used in part or inwhole for media or presentation purposes, on the strict understanding that SDCEA and its copyright are acknowledged. No part of this publication may be reproduced without the prior permission of SDCEA.

Legend

Multi Point Plan Air Monitoring Stations

Roads

- National Roads
- Main Roads
- Other Roads
- Polluting Industries
- Rivers
- Placenames
- Indian Ocean

The companies shown as 'Polluting Industries' were extracted from the South Durban Emission Inventory, which lists emitters of Sulpur Dioxide and other criteria pollutants.

PART TWO:

DURBAN COP 17 CONFLICT

Chapter 6

The Politics of Climate Change in South Africa

Del Weston

In a recent invitation to civil society, the South African government stated that it 'regards climate change as one of the greatest threats to sustainable development'. It continued: '[w]e also believe that climate change, if unmitigated, has the potential to undo or undermine many of the positive advances made in meeting South Africa's own development goals'.[1]

Such rhetoric and South Africa's climate change policies will come under the spotlight during the United Nations COP 17 Climate Summit being held in Durban in late 2011. This paper will provide a background vignette for assessing the rhetoric and politics, in particular the World Bank loan to electricity giant Eskom in 2010, to illustrate the real direction the South African government is taking with regard to climate change.

As an aspiring political and economic hegemon on the African continent, South Africa's climate change politics are of paramount importance, not only because of the country's high carbon emission levels, but because of the power and influence South Africa has over the development agendas of other African nations. South Africa exemplifies the contradictions between capitalism and ecological sustainability. It also graphically illustrates that global warming is but one indicator of the general health of the biosphere and but one of a suite of problems – financial crises, poverty, inequality, urban slums, and social unrest, to list a few – arising from the particular social relations of production under capitalism.

South Africa is also important, because, as Samir Amin (1997) argues, South Africa is

> a kind of microcosm of the world capitalist system, which brings together in a single territory a number of features peculiar to each constituent category

> of that system. It has a white population which, in its lifestyle and standard of living belongs to the "first world", while the urban areas reserved for blacks and coloureds belong to the modern industrial "third world", and the bantustans[2] containing the "tribal" peasantry do not differ from peasant communities in Africa's "fourth world".

South Africa has the second most energy-intensive economy in the world. It is structured around large-scale, energy-intensive industry and mining – the industrial and mining sectors consume over 60 percent of the country's total electricity output (Hallowes and Munnik 2006:25). The economy is underpinned by coal as the primary source of energy. The extraction rate of South Africa's vast reserves is increasing in large part to supply coal-fired electricity for multinational conglomerates such as BHP Billiton, Arcelor Mittal, and Anglo American. This electricity supply is provided under apartheid-era special pricing agreements that provide some of the cheapest electricity in the world for less-than-cost (about one-eighth of what domestic consumers pay [Bond 2011]). Currently, 30 percent of the population have no or minimal access to electricity. Millions of other working class people face disconnections due to dramatic price increases to pay for the construction of two new coal-fired power plants.

Despite both peak oil and global warming, South Africa is expanding its coal-intensive electricity infrastructure, locking the country into a fossil fuel future that contradicts any political claims of attempting to mitigate climate change. Electricity supply is tied to the type of economic structures that are determined by the needs of global capital rather than the needs of the population, a factor too often ignored in the debate around South Africa's energy crisis.

The role of the World Bank

This expansion programme is in part underwritten by a 2010 World Bank loan of US$3.75 billion to Eskom, to enable the company to build the new Medupi Power Station – the fourth largest coal-fired plant in the world – in the Limpopo Province.

Not only does this loan bring further into question the World Bank's role as a major international climate change negotiator and lending agency, it also raises the Bank's tainted history with South Africa. Between 1951 and 1967, at the heart of the apartheid years, it lent US$100 million to Eskom. Victims of apartheid are now seeking reparations in US and European courts from international banks that made money from the profits related to these loans but failed to develop or supply power to any black communities in South Africa. Furthermore, as Bond (2010a) points out, the Bank co-authored the 1996 Growth, Employment and Redistribution (GEAR) macroeconomic programme. GEAR has proved disastrous to the majority; after its implementation, black incomes fell below 1994 levels, while white incomes grew by 24 percent by 2000. Swilling (2010) argues that the Eskom loan gives the World Bank a stronger grip on the South African economy and even greater influence over policy.

Proponents for the loan in the business and political sectors argued that South Africa had no alternative – the country would be plunged into darkness, and the economy would come to a standstill if new coal-powered energy wasn't generated (Jacks 2010). It was argued that the loan was needed to enable universal access to electricity. This argument maintained that the country needed the Medupi coal-fired plant by 2012. Eskom tried to cast Medupi as a green project with the promise that emission-reducing technologies would be built into the plant, which features supercritical air cooling. However, a gigantic system to cool the massive quantities of water that the plant will use is also required in this precipitation-scarce area and will cost more than US$1 billion to construct.

Internationally, more than 140 civil society groups – as well as 60 from South Africa (Bond 2010a) – criticised Eskom's loan application. Hallowes (2009) observes that with regard to African fossil fuel development '80 percent of projects that the World Bank invested in between 1992 and 2003 were designed to export oil to Western Europe, Canada, the US, Australia, New Zealand and Japan' and that, coupled with this 'South Africa's energy- and carbon-intensive export-led economy, an economy in which 80 percent of energy is used by industry and commerce, belies the argument that the loan has anything to do with universal access to energy by domestic consumers'.

In effect, South Africa's poor majority was locked into a debt they did not want but will have to pay for in the short term through increased electricity prices, and in the longer term, through declining environmental conditions and, eventually, the ramifications of Medupi's carbon emissions. With the very explicit support of the World Bank, economic development continues in a manner contrary to what is needed to address the needs of the poor majority, or to address global warming. The argument that South Africa has carbon emission space because it has large-scale poverty to overcome is a red herring – a case of the rich and powerful hiding behind the poor. According to Baker (2011:6),

> Referred to by the government as "Africa's first clean coal 'supercritical' power station", when complete [Medupi] will emit approximately 30 million tonnes [metric tons] of CO_2 per year.

Another dimension of concern regarding the World Bank loan is that the ANC will benefit financially from the construction of Medupi (through its Chancellor House investment in Hitachi Power Africa) and hence from the loan. Hitachi Africa is 25 percent owned by the ANC's investment arm, Chancellor House and, as a partner in the consortium constructing the Medupi power station, was awarded a R38 billion (around US$5 billion) Eskom contract (Bond 2010b). When former Eskom chairman Valli Moosa made the decision to award the R38 billion Eskom contract to Hitachi for the building of Medupi, Moosa was a member of the ANC national executive committee, a conflict of interest deemed 'improper' by a state investigator. Opposition party leader Helen Zille (*Mail & Guardian* 2010) claimed that the ANC will make an estimated R1 billion from the building of the power station, 'enough to fund its election campaigns and the lifestyles of its leaders for years to come', although other ANC sources estimate the profit at closer to R50 million. Zille claims 'no opposition party may ever be in a position to compete fairly with the ANC again' (Boyle 2010). The ANC will entrench its single-party dominance and, in doing so, weaken South African democracy.

The World Bank loan is an illustration of a lost opportunity by the South African government to invest in more equitable, environmentally benign, and global warming-mitigating energy infrastructure.

The new Medupi power station will need enormous amounts of coal, and thus encourage more coal mines. One area which is being more intensively surveyed and mined for coal is in the Limpopo Province. It is here that the Exxaro company has the world's largest open cast mega-mine – a 3.2 kilometre wide, 110 metre deep swathe that is going to be 'driven all the way to the Botswana border, where it will meet up with another open cast mine' (Schalk 2011). The impact already is a vast black, tailings-strewn, toxic, lifeless, and unrecoverable wasteland – another heat-absorbing tipping point for global warming.

The South African government's support for the expansion of coal-fired electricity is driving a new 'coal rush', with an increasing number of mining companies applying for coal mining licences. An Australian company, Coal of Africa, is one player in this new 'coal rush' in the Limpopo Province. Coal of Africa has been developing the Vele coal mine on an 8 000-acre site just seven kilometres east of the Mapungubwe

World Heritage Site and National Park boundary. The Vele site will involve both underground and open cast mining and is being contested by local community groups and NGOs.

However, expansion is already on the agenda with prospecting now occurring on farms throughout the surrounding area (Hatty 2009). Some farmers are dealing with the threat of mining by selling their properties. Entities such as the Black Economic Empowerment (BEE)[3] partner, Mvelaphanda Group, which was previously headed by Tokyo Sexwale (a former high-level ANC member before he was appointed to the Cabinet of the South African government) have partnered with Coal of Africa at various times. Opposition to Coal of Africa's activities remains in the courts. However, this struggle is being acted out on many stages – both micro and macro – across the globe around the competing forces of economic growth and the accumulation of profits in an environment dominated by the imperatives of neoliberal capitalism.

As Coal of Africa illustrates, such companies have significant resources and influential networks to draw upon to win their objectives, which trump concerns over ecological health and integrity, social upheaval and issues of poverty and inequality, history, and culture. Community opponents to such developments are often poorly resourced, fragmented, and not politically connected. More importantly, though, in South Africa the corporations have won over the state to their corner. Regulation is no answer when the rules of the game are set by the corporate captured state.

It is no coincidence that South Africa has adopted inappropriate technologies, such as coal, nuclear, and genetically modified[4] crops – all technologies most suitable for capital accumulation. They all have embedded in them the sense of prestige, power and control that comes with large-scale, sophisticated, and science-manipulated technologies in contrast to the smaller-scale, more democratic and accessible technologies (such as solar and wind power, and local community-based food production), that engender diversity, resilience, improvisation, innovation, self-sufficiency, equity, and justice.

Hallowes (2009:28) concludes that the World Bank loan and Eskom's government-endorsed expansion programme represent a case of 'recreating the logic of an economy that is internally subordinated to the interests of the minerals and energy complex and externally subordinated to the imperial market'. It effectively enables South Africa to export cheap electricity while suffering a shortage of affordable electricity for ordinary people at home.

The building of further aluminium smelters, growing the economy through attracting an energy-intensive and highly polluting beneficiation industry and continuing to supply cheap or underpriced electricity to the minerals energy complex (MEC), is a very short-sighted and irresponsible road to economic development, even without global warming. With global warming, it is a tragedy.

Coal, electricity, and the economy

The South African economic and political elite are key players at the centre of capitalism in sub-Saharan Africa (Bond 2003; McDonald 2009). For example, the New Partnership for Africa's Development (NEPAD) was a neoliberal initiative of the Mbeki government in 2001. And the spread of a centralised fossil fuel-based energy system throughout southern Africa continues, parallel to the exploitation of cheap and free labour from surrounding countries (McDonald 2009).

Many of the ANC political elite are now entwined with the powerful economic interests in South Africa and beyond on the continent. They have become integral players in the neoliberal expansionist project, as well as being intimately involved at a personal level in patronage, personal benefit, and corruption. As Franz Fanon warned[5], the dangers of co-option of revolutionary goals by (neoliberal) capitalist policies amongst those ushering in the new order are real. This is precisely what has happened in South Africa. Here the capital accumulation imperative, and its impact on global warming, are perhaps best described by István Mészáros (2001), when he spoke of capital's 'destructive uncontrollability'.

The South African state is dominated by particular class and economic imperatives that revolve around capital accumulation, growth, and prestige. As Fine and Rustomjee (1996) argue, paradoxically in South Africa, the same class interests can potentially be served as much by nationalisation as by privatisation. This observation is particularly relevant in the South African economy and is a reflection of the merging of private and state capital – and the capture and corruption of government by leading, mostly ANC, ministers and officials. As Grobler (2009:10) writes:

> the Minister for Education's husband has a mining company; the Minister for Transport's wife has a mining company; the Minister for Water Affairs, who used to be the Minister of Minerals and Energy's husband, has a mining company. The ex-Director-General of Trade and Industry and four of South

Africa's provinces' previous premiers own or are BEE partners in mining companies. The list goes on and on.

These elites have been corrupted as beneficiaries of the corporate sector – often, but not only, through the institution of Black Economic Empowerment. At the same time, they are cut off from and indifferent to the people they represent. They are able to and do use the state and state-owned assets to promote their own material interests through networks of political and economic patronage, nepotism, and corruption. Although this is recognised publicly as corruption, there is, as yet, no body with the power and political will to put an end to it. It is claimed (see for example Persson et al. 2010) that no one can expose another's corrupt practices in high offices, because everyone has something on someone else. Multinational corporations that do business in this environment are a central part of this corruption, as is the World Bank in making its recent loan.

This can be seen perhaps as an inevitable part of the process of capital accumulation and particularly neoliberalism, which facilitates illicit capital flows and a deregulation of the financial part of the economy. It can also be argued that corruption arises from an economic doctrine that offers no moral framework, no higher purpose other than to accumulate. At the centre of this accumulation of capital and wealth are the systemic linkages between the political elite, mining, energy, and finance sectors, which arise in two ways: first through investments, acquisitions and mergers, and second through the movement of people in top management positions through parastatal institutions such as Eskom, and private MEC/financial entities.

The phenomenon of Black Economic Empowerment has facilitated these linkages and the flow of benefits into (black) private hands – a new development only in the sense of the inclusion of the black elite. BEE has facilitated the structural expansion of the white elite class to now include a small, black elite, overcoming the international constraints to capital accumulation posed by apartheid, while maintaining the basic class structure of South African society.

The economic and political powers involved around the coal and electricity sectors directly impinge on the environmental concerns about global warming, which span the gamut of immediately tangible environmental, governance and social issues – food and water security, health, air pollution, and socio-economic issues of electricity access, poverty, and inequality. They also threaten the young and already fragile democracy in South Africa. It is not just that the coal and electricity sectors are integral to the minerals energy complex, and as such at the centre of the nation's macroeconomic policies, but they are to a considerable extent driving macroeconomic policy through the enormous power wielded by corporations such as Anglo American and BHP Billiton.

The deregulation and financialisation of the globalised economy – a structural reality that enables the corporations to move their capital and profits rapidly in and out of countries according to where the greatest profits can be made – can only have strengthened corporate power and influence. This is a deeply disadvantageous environment for any national developmental model that is to benefit the people. It is not the corruption of the political elite that is the fundamental problem. Rather it is the fundamental social relations of production, the ensuing construction of the state and the financialised global political economy – which are determining South Africa's and the world's future. With regard to the former, as Harvey (2003:91) writes,

> the preferred condition for capitalist activity is a bourgeois state in which market institutions and rules of contract (including those of labour) are legally guaranteed, and where frameworks of regulation are constructed to contain class conflicts and to arbitrate between the claims of different factions of capital.

The power of the large corporations derives in part from the weight of such large-scale capital investment in the sector and the ramifications of that investment for the rest of the economy, but also, because of the state's commitment to the capitalist economic model.

Carbon trading as the capitalist climate 'solution'

What does all of this mean for South Africa, global warming, and the UN COP 17 Climate Summit? It appears that two leading figures, South African Planning Minister Trevor Manuel and Foreign Minister Maite Nkoana-Mashabane, will be promoting the same economic model that has placed South Africa first in the ranking of unequal societies. Manuel and Nkoana-Mashabane will not be supporting an extension of the crux of the Kyoto Protocol – that is, binding emissions cuts – but will be supporting an expansion of carbon trading.

Carbon trading has been widely criticised elsewhere (Lohmann 2007; Morris and Worthington 2010). As is the case elsewhere, placing climate mitigation strategies in the realm of the market in South Africa will likely generate profits for some of the larger corporations without any real action on global warming. The financialisation of carbon emissions is placing the future of the planet in the hands of the corporate, financial, and banking sector. As the Friends of the Earth report *Subprime Carbon? Re-thinking the World's Largest New Derivatives Market* (FOE 2009:2) emphasises: 'the buying and selling of carbon is fundamentally derivatives trading. Currently, most carbon is sold as futures or forward contracts, a type of derivative'.

Even the more highly regulated European Union's emissions trading scheme (ETS), over its five-year life to 2012 will result in carbon savings amounting to less than one-third of 1 percent of total emissions – a miniscule amount that probably has more to do with the economic downturn resulting from the global financial crisis than the ETS itself (Morris and Worthington 2010).

The Friends of the Earth report adds:

> Today speculators do the majority of carbon trading, and they will continue to dominate as carbon markets grow. In fact, about two-thirds of carbon investment funds by volume were not established to help companies comply with carbon caps, but rather for capital gains purposes (FOE 2009:4).

Furthermore, the report claims:

> The financialisation of carbon trading is placing the future of the planet in the realm of an unstable, unregulated, speculative, casino economy which has no end other than to make increasing profits for increasingly small numbers of people.

Such financialisation has no connection to the lives of the great mass of people, other than to siphon off increasing percentages of their share of the world's wealth. Carbon trading is anti-democratic, given the 'international capital markets [ability to] place serious limits on state authorities to regulate their economies' (FOE 2009:87).

The idea that the future of the planet might rest with derivatives is truly alarming. Ferguson (2008) reports: 'In 2002, legendary investor Warren Buffett warned that derivatives were time bombs and "financial weapons of mass destruction" that could harm not only their buyers and sellers, but the whole economic system.' He was proved right. Buffett (2002) also said:

> The derivatives genie is now well out of the bottle ... Central banks and governments have so far found no effective way to control, or even monitor, the risks posed by these contracts. Closing down a derivatives business is easier said than done. The reinsurance and derivatives businesses are similar: Like Hell, both are easy to enter and almost impossible to exit.

There is little evidence to suggest that these derivatives markets are any more controllable now than then. Yet it is this 'time bomb' that is the financial basis of the carbon trading in which so many governments, including South Africa's, want to place their faith to save the planet.

South Africa is facing multiple crises as a result of its development path and the global capitalist project. Global warming will be a significant and growing factor that will interact synergistically with other crises – such as food insecurity, poverty, inequality, drought, and poor governance. Added to this will be the pressures wrought by yet more refugees and political instability in neighbouring countries of sub-Saharan Africa as increasing numbers of people are forced to leave their (ecologically) inhospitable homes. South Africa needs to change course and protect its environment, its heritage and biological diversity; recover and protect its democratic institutions; protect its waterways and land; and support and regenerate its small-scale agriculture and food growing sectors[6]. Foremost, it must protect and support its people, pull down the barbed wire enclosing the rich enclaves, and give the vast numbers of poorly paid, highly exploited security and domestic workers meaningful and productive employment and compensation for past exploitation. South Africa needs to democratise its institutions, de-globalise, de-commodify, nationalise and redistribute its wealth, overthrow its social relations of production, and become a good neighbour on the African continent, supporting sustainable, equitable, and inclusive human development for all.

The coal industry and electricity sectors – as parts of a complex structure straddling the financial, political, mining and energy sectors – have captured the state and are driving an agenda that is contributing to a catastrophic disjuncture between the interests of the majority and those of a small but powerful national and global elite. What emerges in practice is that the coal and electricity sectors are part of an accumulation process that disenfranchises, impoverishes, and further alienates the poor. At the same time, it destroys the broad ecology, which results in declining natural resources for the poor to fall back on for survival.

Conclusion and summary

South Africa illustrates a number of key points that are important to the political economy of global warming.

First, South Africa's ruling political elite is as deeply committed as its predecessors to the process of capital accumulation. This is obvious even in relation to the coal mining industry and coal-fuelled electricity energy sector, despite the costs to the people, the environment, the incontrovertible growth in carbon emissions, and the resulting difficulties it will create in adapting to global warming.

Second, the financialisation of the globalised economy and the internationalisation of corporate activity have exacerbated both class divisions and the rift between society and the ecology, as the rush to mine coal and other minerals continues and the profits from such activities leave the country.

Third, international institutions such as the World Bank play a pivotal role in the endorsement of this political economy.

Fourth, South Africa illustrates how the coal and electricity sectors are integral to the sub-imperial expansion of capitalism throughout sub-Saharan Africa. Further, it shows that the coal sector is central to the process of accumulation by dispossession – the dispossession of communities from their land, and the equally insidious dispossession of people from their livelihoods and health through the poisoning of the land, the water, and the air. The Medupi Power Station – with its World Bank loan, its projected emissions, its *raison d'être* being cheap power for multinational conglomerates, its corrupt links to the ANC elite, and its need for yet more coal mines – is an excellent example of the anti-social, anti-ecological contradictions of neoliberal capitalism in South Africa.

Fifth, the particular social relations of production unique to mining sites around the world and especially in Third World and middle-income countries where labour is cheap, expendable and inadequately protected – necessary conditions for accumulation and multinational investments – have been causally linked to health problems such as tuberculosis and silicosis. The use of migrant and internal cheap labour, dislocated from their families and communities, have contributed to the HIV/AIDS epidemic. This is the ultimate exploitation and alienation, when people lose their health and lives due to the particular social relations of production, while the political elite seeks to keep up the façade that they are working to raise the masses out of poverty. At the same time the marketisation of the economy within South Africa and beyond – including land grabs and the industrialisation of farming – transforms increasing numbers of rural small holders into the proletarianised unemployed without a future – again destroying people, communities, and the ecology.

Sixth, the model of development has not only further impoverished the poor but widened and deepened the gap between rich and poor. It is continuing to undermine the health of the broad ecology that the poor depend on, and will increasingly depend upon, for their survival. Related to this, the IPCC (2007) has stressed the importance of good governance and infrastructure, equity, sustainable economic development, and carbon emissions reductions as essential ingredients to cope with global warming. South Africa is undermining and destroying these vitally important environmental and social lifelines for the future. If one throws into this mix a growing refugee population, internal racial tensions and xenophobia, declining food and water security, an extraordinary lack of moral leadership and political will to find a better way – as well as global warming – the future of South Africa looks bleak indeed.

Seventh, another by-product of neoliberal capitalism is the predominance of money and the market. The incubators of power, greed, and corruption have destroyed the integrity of the ANC and contributed to the skewing of the economy toward sectors that make money for a few, and where corruption greases political policies and political careers.

Finally, while there are manifold demonstrations against the poor conditions under which people live, the left social movements and trade unions are the main actors in South African society consistently opposing this form of development. The uprisings in Northern Africa and the Middle East since early 2011 are beacons of hope and inspiration for oppressed people on the continent. The factor of global warming, however, leaves only a small historically-specific and short window of opportunity for change. Otherwise for South Africa, in five to ten years, the political, economic, social and ecological prospects are grim as the impacts of global warming are more dramatically manifested. At such a point, it will become increasingly difficult and perhaps too late to carry out any sort of peaceful, transformative structural change.

The capitalist development model with its inexorable dependence on fossil fuels and relentless pursuit of growth and expansion stands in complete contradiction to the requirements of the biosphere and humanity. South Africa exhibits the consequences of a state dominated by the interests of conglomerate capital, with all the resulting aspects of the deepening rift between humans and the environment. It exposes the commonalities of interests of global elites, the merging of corporate and state interests, and the significance of power and hegemony in preventing the development of metabolically restorative and harmonious communities. It demonstrates the distinctive, hegemonic and dangerous power of corporations to society.

While South Africa can be seen as a microcosm of the global economy, it is also one country that will feel the impacts of global warming catastrophically. Under current practices, predictions are that by the end of this century, rising temperatures will reduce the capacity of the planet to support a human population of no greater than 1 billion people – the loss being exponential and bunched toward the end of the century (Spratt 2011). The effects of global warming are predicted to most severely impact Africa. These are points to consider in reviewing South Africa's official political stance at the COP 17 Climate Summit.

References

Abalimi Bezekhaya Newsletter. No 36. April 2008–September 2009, http://abalimi.org.za/newsletters/nl36.pdf, accessed March 2011.

Amin, S. 1997. Samir Amin's 1997 Babu Memorial Lecture, http://www.pambazuka.org/en/issue/451, accessed March 2011.

Baker, L. 2011. Governing electricity in South Africa: wind, coal and power struggles. Working Paper 015, The Governance of Clean Development. Norwich: University of East Anglia.

Black Economic Empowerment (BEE) Report. 2010. Black Economic Empowerment Commission. Johannesburg: Skotaville Press.

Bond, P. 2003. *Against global apartheid. South Africa meets the World Bank, IMF and international finance*. Cape Town: UCT Press.

———. 2010a. South Africa's bubble meets boiling urban social protest. *Monthly Review* 62 (2).

———. 2010b. Eye on civil society. The bank loan that could break SA's back. 13 April. *Monthly Review*, mrzine.monthlyreview.org/2010/bond130410.html, accessed March 2011.

———. 2011. Introduction: Welcome to Durban. *Capitalism Nature Socialism* 22 (4).

Boyle, B. 2010. Eskom loan must not benefit ANC. April 4.

Buffett, W. 2002. Warren Buffett on derivatives. Excerpts from the Berkshire Hathaway annual report for 2002, http://www.fintools.com/docs/Warren%20Buffet%20on%20Derivatives.pdf, accessed June 2010.

Ferguson, A. 2008. Economies count the cost of derivatives. *The Australian*, http://www.theaustralian.com.au/business/opinion/counting-cost-of-derivatives/story-e6frg9j6-1111117783464, accessed March 2011.

Fine, B. and Z. Rustomjee.1996. *The Political Economy of South Africa. From Minerals-Energy Complex to Industrialisation*. London: Hurst & Company.

Friends of the Earth. 2009. *Subprime carbon? Re-thinking the world's largest new derivatives market*, http://www.foe.org.au/media-releases/2009-media-release/carbon-trading-the-next-sub-prime-new-report/, accessed March 2011.

Fig, D. 2009. Nuclear energy rethink? The rise and demise of South Africa's Pebble Bed Modular Reactor. Institute for Security Studies, http://www.iss.co.za/uploads/210.pdf, accessed September 2011.

Grobler, J. 2009. Testing democracy: which way is South Africa going? http://www.scribd.com/doc/31603161/State-of-Democracy-in-South-Africa, accessed November 2009.

Hallowes, D. 2009. *The World Bank and Eskom: Banking on climate destruction*. Pietermaritzburg: groundwork, http://www.groundwork.org.za/.../14Dec09EskomReport.asp, accessed April 2011.

Hallowes, D. and V. Munnik. 2006. *Poisoned spaces: Manufacturing wealth, producing poverty*. October. Pietermaritzburg: groundwork, www.groundwork.org.za/Publications/groundWorkReport2006.pdf, accessed March 2011. 25.

Harvey, D. 2003. *The new imperialism*. Oxford: Oxford University Press.

Hatty, P. 2009. Mopane Bush Lodge, Limpopo Province. Personal Communication.

IPCC. 2007. Pachauri, R.K. and A. Reisinger, eds. *Contribution of Working Groups I, II and III to the Fourth Assessment Report of the Intergovernmental Panel on Climate Change*. Geneva: IPCC.

Jacks, M. 2010. Analysts sneer at coal-fired concerns. Eskom will use very clean technology say economists. *Johannesburg, Business Report*, http://www.busrep.co.za/index.php?fArticleId=5410040, accessed April 2011.

Lohmann, L. 2007. *Conversation on climate change, privatisation and power*. Development Dialogue, Dag Hammarskjold Foundation, No. 48. Durban: The Durban Group for Climate Justice and the Corner House.

Mail & Guardian. 2010. Zille says ANC stands to make R1bn from Medupi. 3 April, http://www.mg.co.za/article/2010-04-03-zille-says-anc-stands-to-make-r1bn-from-medupi, accessed March 2011.

McDonald, D.A. 2009. The importance of being electric. In *Electric capitalism. recolonising Africa on the power grid*, ed. D.A. McDonald. Cape Town: Earthscan: 15–23.

Mészáros, I. 2001. Socialism or barbarism. *Monthly Review*, http://www.monthlyreview.org/1201jbf.htm, accessed March 2011.

Morris, D. and B. Worthington. 2010. Cap or Trap. How the E.U. ETS risks locking-in carbon emissions. London: Sandbag, www.sandbag.org.uk/.../2010/.../eu-emissions-trading-scheme-threatens-trap, accessed March 2011.

Persson, A., B. Rothstein, et al. 2010. *The Failure of Anti-Corruption Policies. A Theoretical Mischaracterization of the Problem*. QoG Working Paper Series 2010: 19, University of Gothenburg, http://www.qog.pol.gu.se/working_papers/2010_19_Persson_Rothstein_Teorell.pdf, accessed April 2011.

Schalk, B. 2011. Expansion on track to deliver coal to Medupi. *Engineering News*. Johannesburg.

Spratt D. 2011. When the climate breaks bad, 'Will there be answers?' *Dissent* 35: 25–30.

Swilling, M. 2010. Intelligent power: SA is throwing it all away. *Cape Times*: 9.

Endnotes

1 http://cop17insouthafrica.wordpress.com/.

2 Bantustans refers to the territory that was set aside under the apartheid regime for black South Africans.

3 The South African government under the ANC launched the Black Economic Empowerment programme to redress inequalities that the non-white majority suffered under apartheid. According to the BEE Commission Report, BEE 'is an integrated and coherent socio-economic process. It is located within the context of the country's nation transformation programme, namely the RDP (Reconstruction and Development Programme). It is aimed at redressing the imbalances of the past by seeking to substantially and equitably transfer and confer ownership, management and control of South Africa's financial and economic resources to the majority of the citizens. It seeks to ensure broader and meaningful participation in the economy by black people to achieve sustainable development and prosperity' (BEE Commission Report 2001:2).

4 South Africa succumbed to pressure from the biotechnology companies and is now the sixth largest producer of genetically modified (GM) crops in the world. In 'the past five years, hundreds of thousands of hectares have been turned over to planting of GM crops ... Small farmers have been supported to plant GM cotton ... but are increasingly finding difficulties with debt, low cotton prices, withdrawn extension of services and a lack of irrigation' (Fig 2009).

5 From a quote from Lewis Gordon, Professor of African Studies, in *Monthly Review*, July 2002, in a book review of Ashwin Desai's *We Are the Poors: Community Struggles in Post-Apartheid South Africa*.

6 South Africa used to grow approximately 40 percent of its food; it now produces about 25 percent (Abalimi Bezekhaya Newsletter 2008–2009).

Chapter 7

Colonising Africa's Atmospheric Commons

Khadija Sharife

At the January 2010 Investor Summit on Climate Risk conference held in New York, more than 450 investors controlling over US$13 trillion in assets declared that action must be taken to pre-empt international climate change treaties, in order to develop sustainable economies – chiefly through the carbon market. 'Copenhagen was a missed opportunity to create one fully-functional international carbon market,' conceded Peter Dunsombe (UNEPFI 2010), head of the Institutional Investors Group on Climate Change, which comprises European financiers.

The United Nations Environment Programme anticipates that 85 percent of the finance required to make the shift will be derived from private investors. The commodification of pollution is inspired by the rationale of market efficiency: major polluters issued with permits are incentivised to emit less, thereby enabling them to make a profit selling excess permits to those less efficient. In order to limit the pollution bubble, the Kyoto Protocol's 'flexibility mechanisms' facilitate a process allowing for polluters to finance carbon-light projects in countries that would otherwise engage in conventional methods. By doing so, securities are generated through various 'offset' tentacles designed to exploit the *underdeveloped* status of countries that fail to access and utilise their share of the atmospheric commons.

As a result, it is worth asking whether the Kyoto Protocol should be extended, as is, for a second period beyond the current emissions target cut-off date in 2012. There is certainly a need for a legally binding deal that makes substantive emissions cuts, so for that reason there is a strong progressive lobby – led by the Third World Network – to extend Kyoto. But there is a downside – especially evident in Durban: thanks to offsets that grant industrial polluters a get-out-of-jail-free card through Kyoto's Clean Development Mechanism (CDM), polluters circumvent caps by investing in greenwashed business-as-usual projects in poorer countries. Offsets such as Durban's methane-electricity conversion CDM at the Bisasar Road landfill site, or timber-related offsets emerging across the continent, facilitate the process of corporate gaming, simultaneously utilising Africa's underdeveloped status as yet another exploitable resource, with the added danger of 'enclosure' threats to residents.

Bisasar Road, Durban

Twenty-five years ago, in my old home neighbourhood of Clare Estate in the relatively close-in suburbs of Durban, a vast dump was designated in what had been a pristine valley. It would become Africa's largest formal landfill, and was simply imposed upon a black community (specifically with 'Indian' and 'Coloured' race classifications, later to be joined by African informal residents known as shack dwellers). In 1994, there were widespread written promises of Bisasar Road's closure by Nelson Mandela's African National Congress. Delays occurred as other landfill closures in a poor black neighbourhood (Umlazi) and Umhlanga – the wealthy and upmarket area, characterised then by majority white ownership and capital – were prioritised. Then, because of the World Bank's interest in carbon trading, peaking in 2002, and thanks to the Kyoto Protocol, a decision was made by municipal officials to leave Bisasar Road open for business.

The area is known for its carcinogenic atmosphere, where seven out of ten households on the western side have 'reported tumour cases' (Reddy 2005). Community activist Sajida Khan died of cancer in 2007 fighting the dump. She diagnosed the larger political cancer: 'Poor countries are so poor they will accept crumbs. The World Bank knows this and they are taking advantage of it' (Bond 2007).

As a source of electricity, the project is more symbolic than serious, generating an insignificant 6 MW of so-called *green power electricity* from methane mixed with toxic gases; Durban's daily consumption is around 1 500 MW. But the damage of maintaining the dump so as to extract methane for many more years is that medical, industrial, and domestic waste continues to poison the land and water, with methane seeping into surrounding homes. An alternative advocated by Khan was to close the dump, install collection pipes, and integrate the landfill gas flow (7 000 cubic metres per hour) to Petronet's pipelines running past the site. This idea was rejected by

Durban's leading official involved with the project, who stated, '[What] if something goes wrong with this pipeline? [What] [i]f the land subsides or they do something funny with their pipeline? Don't send the problem to someone else's backyard and tell them to sort out our methane. We think it's very irresponsible. The world thinks it's very irresponsible' (Reddy 2005).

According to John Parkin, deputy head of engineering at Durban Solid Waste, 'What makes them worthwhile is the revenue that can be earned from carbon credits.' Bisasar's CDM classification neatly legitimises the presence and active state of the lethal and carcinogenic dump for US$15 million in certified emissions reductions (CER) credits for methane-to-electricity conversion over a 20-year lifespan. Trevor Manuel, previously the country's minister of finance and now head of a new planning ministry, recently stated that such projects would potentially be the basis for revenue to the Green Climate Fund (GCF) up to half its anticipated US$100 billion per year, and be subject to tax exemptions. In March, Manuel was made co-chair of the GCF design team, and in July he repelled calls from Nicaragua's UN delegation to disqualify the World Bank from the team on grounds of conflict of interest, since they are interim GCF trustees (Bond and Sharife 2011).

Extending the African resource curse to forests

African states are great fans of tax 'competition': a large portion of the continent's exports (80 percent of which are primary commodities) are negotiated via secretive development agreements, swapping corporate tax holidays (the main source of income from rents) for bribes. It is precisely the lack of revenue disclosure, redistribution, and re-pricing – in the process, concealing socio-ecological costs of extractive industries, which are classified as externalities or hidden costs – that has catapulted Africa into the position of the continent most vulnerable to climate change, according to the UN's Intergovernmental Panel on Climate Change (IPCC) (UNEP 2006). Unlike developed nations, which derive 80 percent of their wealth from intangible capital (such as education), African citizens depend on the health and evolution of direct ecosystem services such as fisheries, fertile land, and fresh water, which account for 70–80 percent of the continent's natural wealth (*African Business* 2009).

This political economy is the opposite of Africa's concept of *Ubuntu*: people are people through each other's humanity. But humanity cannot sustain itself without the resources that, under capitalism, must be legally recognised and commodified before development (via sustainable economics) can take place. The result: a resource-rich, mal-developed continent controlled through forced peace – i.e. electoral authoritarianism or militarisation. Though Africa remains a major supplier of fossil fuels, over 30 percent of the total human population living in darkness resides in Africa, with the continent contributing just 5 percent to global electrification and 3.5 percent of carbon emissions (UNEP 2006).

Paradoxically, Africa supplies as yet legally unrecognised ecological services to the developed world, such as the Central African rainforest. These forests – of which one-third are found on the African continent — cumulatively absorb a figure almost equivalent to the ecological reparations offered by developed nations for exhausting the atmospheric commons i.e. £13 billion (*Science Daily* 2009).

But global warming, packaged as an apolitical environmental issue, has yet to be placed in the context of 'selective justice.' The ecological debt from rich nations to poor nations is still notional, even if various studies measure it in the tens of trillions of dollars – a fact that even media vanguards of the developed nations, such as CNN, have acknowledged, citing the US$2.3 trillion 'eco-debt' ecological economists at the University of California, Berkeley claim is owed to Third World countries (Oliver 2008). Yet the architects of the cap and trade system deliberately blur the lines between victim and aggressor. The former suffer the threat of caps on survival carbon necessary for basic electrification, and the latter seek to enforce ownership of atmospheric rights so as to preserve the fossil fuel intensity of their economies.

One revealing example of the way carbon is being colonised is REDD: Reducing Emissions from Deforestation and Forest Degradation. The initiative is designed to protect and conserve the world's remaining lungs and carbon sinks – forests – where ongoing deforestation and degradation currently account for 17 percent of global emissions from stored carbon. Success will be achieved from stopping these destructive processes, which are taking place primarily in countries that are under-resourced, punctuated by corrupt or diminished states, and unable to police or protect forested land from illegal logging. The REDD initiative also intends to finance the protection and conservation of said lungs: a fifth of the world's fossil fuel emissions are absorbed by forests, with Africa acting as a sink for 1.2 billion metric tons of CO_2 annually (*Science Daily* 2009).

REDD was first proposed in 2005, at the 11th Conference of the Parties (COP 11) by the Coalition for Rainforest Nations, a 15-member group of forested developing countries, including Nigeria, Equatorial Guinea, Liberia, Lesotho, Kenya, Indonesia, and Madagascar. The coalition's self-described goals are to generate revenue streams derived from a programme of 'forest stewardship reconciled with economic development' that is chiefly driven by communities. Multilateral drivers of REDD include the UN-REDD programme and World Bank Forest Carbon Partnership Facility. The Bank is a key financier with a US$300 million fund (Global Witness 2008).

The incentive for African states to receive funding via carbon credits by 'farming forests' appears to be compelling from both financial and ecological points of view. On the surface, the REDD deal offers a double save: firstly, by ensuring that forests remain intact and secondly, by attaching the ability to generate revenue to conservation of forest lands. For instance, the Central African rainforest has the world's second largest forest cover, accounting for 18 percent of the earth's total forest lands. Kenya's 400 000-hectare Mau Forest Complex – East Africa's primary water catchment area – for instance, may average 160 tons of carbon per hectare (Vidal 2009). However, by 'farming carbon', the commodification and financialisation of ecosystem services within the context of capitalist economy occurs – with all of the accompanying socio-economic injustices, ranging from land grabbing to displacement of indigenous peoples.

After 16 years in power, and a heavily disputed election in 2005, Ethiopia's dictator, Meles Zenawi, is at the helm of a country facing severe ecological crises due to mass deforestation caused by illegal logging. The country's Agricultural Research Institute reports a loss of 200 000 hectares per annum (Ababa 2007). The head of Ethiopia's Institute of Forestry Development, Dr Alemu Gezahegn, revealed that Ethiopia would lose all forested land by 2020 if deforestation continued at the current 'alarming pace'.

France itself maintains an extensive logging footprint in former African colonies and other *Francafrique* territories, such as Cameroon and the Democratic Republic of the Congo. Cameroon is one of the world's top five wood exporting countries and hence, its trade economy is chiefly dominated by a small handful of French logging companies such as Coron, Rougier, and Thanry. In 2005, *Inter-Press Service* quoted a senior official at the Cameroonian Centre for Environment and Development based in Yaoundé, who said that NGOs could not name the logging companies for 'fear of reprisal', while 'the police shy away from investigating the matter as well … because those who are profiting illegally from logging allegedly include senior police officials' (Tetchiada 2005). As one French national involved in the logging industry revealed to *IPS*, 'We're asked for bribes amounting to millions of CFA francs, and we often pay these out.'

Logging is big money. In 2003, for instance, Greenpeace (2003) claimed 'using GPS co-ordinates, our investigations have provided evidence of an illegal road network, log ponds and abandoned logs over an area of about 14 square kilometres.

Preliminary estimates indicate that the illegally produced timber is worth about 1.5 million euros.' Though wood from Africa and Asia is increasingly treated and finished in China, Europe remains the primary market. Illegal logging of forested lands, generally termed either as common property resources (thereby indicating government ownership), or as customary or community ownership resources, which may lack ownership altogether, has rendered barren millions of hectares within the Mau Forest Complex in Kenya and across the continent. Sudan, for instance, has experienced the loss of more than 8.8 million hectares of forest; the Democratic Republic of the Congo, 6.9 million ha; Tanzania, 6.2 million ha; Nigeria, 6.1 million ha; and Cameroon, 3.3 million ha.

Ironically, REDD's process is capital intensive, allegedly requiring an average of US$2 000 for every hectare certified after ownership has been legally proved. This renders the process of establishing carbon farming projects similar to other enclave capital intensive industries where states tend to lack the funds required to finance the investment, thus paving the way for foreign financiers. And regimes, whether corrupt or democratic, automatically remain on the receiving end of profit, as long as these forests remain open to investment designed to cash in on pollution, as well as circumvent emission reductions embedded through gaps such as the flexibility mechanisms in the Kyoto Protocol.

As market-maker Ken Newcombe stated at 2004's Carbon Expo in Cologne, 'The World Bank is reducing the risk for private investors' (Lang 2004). And for private investors, the opportunity is tempting. At the Rukinga Ranch in Kenya, for example, wealthy western dotcom entrepreneur Mike Korchinsky and his partner Bob Dodwell spent more than US$400 000 over a period of six months certifying and analysing the 80 000 acres of land they purchased for US$10 per acre, engineered as a deal that would benefit from the REDD scheme. They can expect well over US$2 million in returns annually (Vidal 2009). Chunks of Kenya's Mau Forest Complex have been acquired by bogus companies related to the state with concessions large and small, such as the Moi-connected Sian Enterprises (*Daily Nation* 2009). Others include Olalarusi Inv Far (9 887 acres), the Catholic Church of St. Francis (7 305 acres), Ilgina Contractors (3 202 acres), and the Kiptagich Tea Estate (Ndungu Commission of Inquiry 2004).

Ironically, many like Ilgina, whose directorship is comprised of the powerful Ntutu family (Agnes Naropil Ntutu, Kiteleiki Ntutu, and Kunini Ole Ntutu), were party to the registration and allocation of land via the Ntutu Presidential Commission. In 1986, the Commission demarcated the boundaries of the Maasai Mau Forest. According to Nairobi's *Nation* newspaper, 'members of a powerful [Ntutu] family in Maasai amassed chunks of land, virtually owning the entire Maasai Mau Trust Land Forest in Narok'

(Wafula 2009). Unlike Korchinsky and Dodwell's plan at Rukinga Ranch, where 50 community shareholders will receive returns from the project, and US$600 000 will be ploughed back into the protection of forests (allegedly), the bulk of forest concessions are afforded little, if any, accountability. As Dodwell told *The Guardian*,

> Logging companies may turn into carbon companies. In most countries in Africa you can do what you like, log out the trees, put in roads, do anything. There is little or no monitoring. The rewards could be 99 percent for me and 0.5 percent for the communities (Vidal 2009).

But for the Mau Forest Complex's Ogiek peoples, who were marginalised from ancestral lands during the days of the British Empire, the conservation practised at Rukinga Ranch amounts to nothing more than criminality. One result is the forced displacement of more than 1 650 families since November 2009 (Lang 2009). Even though he supposedly offers the leading hope for democracy in Kenya, Prime Minister Raila Odinga in 2009 suggested that every single Ogiek would face arrest if they did not voluntarily move as part of the government's plan to 'reclaim' the Mau Forest Complex (Lang 2009). This move had been promoted as part of the agenda to secure the Mau's crucial forested land, which also generates East Africa's primary water catchment area that supplies major rivers and lake systems, including the Nile and Lake Victoria, and feeds into Uganda, Tanzania, Somalia, Ethiopia, and Sudan.

Policies resulting in the displacement of vulnerable peoples like the Ogiek mark the general trend of REDD projects. Of 144 projects assessed by the International Institute for Environment and Development, just one project 'included a proposal to make community-managed forests or indigenous peoples' rights a binding part of REDD' (Vidal 2009). And despite peoples such as the Ogiek possessing the complex knowledge base required to monitor and protect the Mau Forest Complex, this cannot be done without according legal rights to indigenous peoples, who occupy such land through customary and community ownership. These are the people branded by the Kenyan government as squatters. According to the Washington-based Rights and Resources Institute, the process would cost just US$3.50 per hectare (Ostrom 2009). But the 'paper parks' backed by the UN have failed to acknowledge forests as *socio-ecological* ecosystems, preferring instead to protect 'natural' land devoid – or cleansed – of peoples, leading to the rationales of the conservation and privatisation tradition.

Norway's REDD investments in Tanzania

Norway-based Green Resources Ltd's plantations are a good case study. As Africa's largest forestation company, at first blush it appears to be an entity with a sustainable mission, ranging from recuperation of wood waste to carbon sequestration. This is no accident: just two days after the UN Framework Convention on Climate Change's (UNFCCC) Kyoto Protocol was adopted, the company (then known as Fjordglott) increased its capitalisation from US$98 000 to US$1.4 million, later extending invitations to private investors, such as Norwegian corporation TRG, to acquire shares (Eraker 2000).

Green Resource's activities include plantations, carbon offsets, forest products, and renewable energy. The company's wood production in Africa is pegged at 14 000 ha of new forest plantations, but a total of 610 000 ha are under process for future development (Green Resources 2009). The company, primarily operating in East Africa with 3 500 employees, holds 12 000 ha in Uganda, as well as significant areas in Tanzania (34 000 ha of land, with a further 120 000 ha in the process of acquisition), Mozambique (172 000 ha), and Sudan (179 000 ha). It also owns East Africa's largest sawmill, Sao Hill, and remains one of the continent's largest producers of transmission poles required for electricity, among other products such as wood for housing. The company's current major shareholders include: Phaunos Timber Fund (11.7 percent), New Africa (30.7 percent), Steinerud (9.3 percent), Macama (7.3 percent), Storebrand ASA (8 percent), Verbene Investment Ltd. (8.2 percent), TRG (4.7 percent), and Preben Invest AS (3.8 percent) (Green Resources 2011).

Unlike the company's limited competitors, Green Resources has the comparative advantage of already having significant experience in land acquisition, the lack of which the company describes as a 'significant entry barrier' (Sharife 2010b). Green Resources' competitors include the Global Forest Solidarity Fund, a private initiative that is funded to the tune of US$100 million from investors such as Harvard University. This concern is active in Mozambique, and has planted 5 000 ha in the last decade. Other competitors include the UK-financed New Forest, which operates in Uganda and Mozambique and which planted 1 500 ha in 2007; Actis/CDC, which controls 7 000 ha of teak plantation in Tanzania, as well as logging rights in Sudan; Raiply, East Africa's largest forest industry company, which owns 12 000 ha in Tanzania and has operations in Kenya; and Rift Valley Holdings, self-described as 'one of the largest investors in agriculture and forestry in sub-Saharan Africa' (Rift Valley 2011).

As Mutuma Marangu, chairman of a Green Resources subsidiary, recently explained in an interview with Canada's *Business News Network*, a tree that takes 70 years to grow in Norway, takes just 17 years in Tanzania. Given the geostrategic location of East Africa and the rising need from emerging nations such as China for wood, Marangu believes that with major forests coming online, Chinese, Japanese, and other major users of wood in the Near and Far East will move toward East Africa as opposed to Brazil, Argentina, and Chile, their traditional mainstays for lumber. 'There is a shorter shipping voyage from China to East Africa rather than China to Brazil,' he stated. 'There is ample opportunity here. We have the least penetration of forestry and forest cover in the world, the greatest possibility for growing trees. We are leading by example.' Marangu added that population growth at 'one million people, per year' in Kenya, Tanzania, and Uganda created an annual demand of wood for 300 000 homes in each of the three countries. 'In terms of power and demand, there is an extreme need for both [wood] production and [carbon] offsets. At present, there is not much multinational or foreign investment in the sector,' he said (*Business News Network* 2010).

Currently, the 'high cost' producers such as the USA (producing over 600 million cubic metres of wood per year) and Russia (which produces over 400 million cubic metres annually) among others such as Uruguay, Brazil, Indonesia, and South Africa, account for 80 percent of global supply (Green Resources 2009). The company noted that Russia's log wood export tariffs at the time (€50 per cubic metre) exceeded Tanzania's stumpage costs (the residual costs after subtracting various allowable costs such as transport).

The optimistic assessments of China and India as consumers are reflected in the doubling of timber imports to both countries between 2002 and 2006. And while China has increased domestic plantation of both hardwood and softwood (rising from 6 million and 7 million cubic metres, respectively, in 2000, to 25 million cubic metres and 16 million cubic metres annually in 2010), China's 'explosion' in wood imports will have only a 'modest impact' (Green Resources 2009).

It is precisely the company's vast wood resources that enable it to perceive carbon offsets as a viable means of generating profits through carbon sequestration or storage via plantations. Plantations are seen as a profitable means of mitigating climate change under the umbrella of the CDM. Despite studies by Stanford University's Program on Energy and Sustainable Development, which reveal that between one-third and two-thirds of CDM projects 'do not represent real carbon reductions' (quoted in Christian Aid 2009), CDM projects account for 20 percent of the official carbon market (Green

Resources 2009), which in 2009 was valued at US$17.5 billion out of US$94 billion. The European Union carbon market comprises 77 percent (US$72 billion), but the market itself is pegged to explode by 300 percent in the event of the US market ever coming online (Macken 2011).

The relevance of forests as a means of *sinking* carbon was recognised by the Kyoto Protocol's Articles 6 and 12, which are related to project activities and emissions trading. Article 6 articulates that Annex 1 (or developed/industrialised nations accounting for more than 80 percent of historical emissions), may transfer carbon credits to or acquire carbon credits from any other Annex 1 country. The result is the creation of various projects aimed at reducing man-made emissions or enhancing carbon sinks (Brown 1999). Included in Article 6 are two key provisions stipulating that any claimed emissions reductions must not be from emissions that would have otherwise occurred. Meanwhile, Article 12 concerns the role of non-Annex 1 nations (developing countries) by defining the role of CDM initiatives, enabling – in theory – developing nations to move forward sustainably through technology transfer (for example, solar and wind power) from developed nations. Afforestation (A) and reforestation (R), or 'A/R', projects were adopted by the 9th Conference of the Parties (COP 9) in December 2003 (Republic of Tanzania 2007).

In 2009, Green Resources declared that existing projects will generate over 60 million metric tons (mt) of carbon capture during the next decade, with total forestation carbon capture expected to peak at 2 million mt per annum for existing projects then under development (Green Resources 2009). Additional projects are estimated to generate 9 million mt by 2020, when 'net growth in biomass is the highest'. The Norwegian government, eager to offset some 6 million carbon credits (TimberWatch 2009) has already acquired carbon credits from Green Resources.

Tanzania ratified the Kyoto Protocol in 2002. The UNFCCC's approval is vital, as is the approval of the host country via the Designated National Authority (DNA). Tanzania's official CDM guide for investors confirmed this structure, revealing that, 'In Tanzania, before the DNA can approve the A/R project activities, it is addressed by the Ministry of Natural Resources and Tourism through the Forest and Beekeeping Division which has formed a task force to look at A/R project activities and other factors related to carbon trade opportunities.'

In order to speed up and simplify the process for small-scale CDM initiatives, the government has implemented 'faster registration, only four weeks after submission, exemption from [a] registration fee,' as well as entities that are 'validated, verified and certified by the same Designated Operational Entities' (DOEs). DOEs are responsible for checking that CDM projects conform to proper regulations. To achieve DNA approval, Project Idea Notes (PINs), which identify the 'additional' nature of the project that enables it to qualify for CDM status, and Project Design Documents (PDDs) are required, while DNA involvement from the project start date is preferred. Projects earmarked for rural areas are also preferred, while technology transfer remains one of four key conditions pending approval by the DNA. The guide further reveals: 'Timing of the CER sales has an impact on the price which can be obtained. Contacting buyers at the PIN or PDD stage can be advisable if financing for the CDM registration is sought.'

Green Resources (2010) describes its Idete Forest Project as one such CDM initiative: 'The objective of Idete is to grow trees for carbon storage and to harvest forestry products for sawn timber, utility poles and renewable energy.' Although the company submitted the PDD in November 2008, planting at Idete had already begun in 2006 on degraded grasslands (Green Resources 2010). Forest Stewardship Council (FSC) certification, already achieved in 2008 at the company's Uchindile and Mapanda concessions (Sharife 2010b), is fundamental to ensuring that best industry standards are maintained. According to Dr Blessing Karumbidza of the University of KwaZulu-Natal, it is Norway – one of the world's leading oil-producing nations – rather than Tanzania that stands to benefit:

> The Idete project was allegedly underwritten by Norway's Ministry of Finance. The Norwegian prime minister, who was present at the time of the launch, articulated the importance of carbon credits as a means of offsetting Norway's emissions. Tanzania was at the heart of the deal (Sharife 2010b).

According to Karumbidza, who also represents the African civil society organisation TimberWatch, the irony is that wood plantations are not forests but monocultures and thus should not receive FSC certification.

> Green Resources claimed that the acquired Idete land was degraded through fire, but unlike wood plantations where fire can destroy wood products, grassland fires serve a very natural and quick process of maintaining the ecosystem by removing dead herbaceous materials, recycling nutrients, and other important factors. The purpose of this and similar CDM deals is not to transfer technology enabling sustainable development and renewable energy in developing countries in exchange for offsetting Western emissions, but instead, more of the same exploitation (Sharife 2010a).

Species are primarily composed of potentially invasive eucalyptus (59 percent) and pine (40 percent). The forest, situated in the Mufindi district of the Iringa region, is located at an altitude of between 1 100 metres and 1 550 metres. The rainy season extends from November through May. By 2008, 1 600 ha of an 8 000 hectare plantable area (from a total of 11 600 ha of Idete land acquired by Green Resources) had been developed, generating a potential 172 471 in temporary certified emission reductions (tCERs) per annum (Green Resources 2008).

The company estimates that total production over 20 years will generate almost 2.6 million tCER from Idete alone. In 2009, Green Resources revealed that potential tCERs of over 6 million by 2020, sold at US$6 per estimated emissions reduction, would generate US$36 million in revenue for the crediting period – a huge ongoing windfall for the company considering that under CDM rules, the accumulated carbon can be sold every five years.

Despite these enormous sums, Green Resources (2008) has described the economic challenges facing carbon offset projects in Tanzania as beset by a 'high level of risk, low and uncertain tCER price, and high cost of project development and implementation'. Institutional and social obstacles include limited government understanding of the carbon certification process and difficulties innate in government procedures and bureaucracies for approval, as well as similarly limited understanding on the part of communities (Green Resources 2008). The company further added in their presentation, 'Overview of Plantation/Certification Development in Tanzania', that stakeholders and communities had very high expectations of perceived benefits. Moreover, the company declared that private investors were placed in a disadvantageous position. 'While there are large amounts of funding available for forestry and carbon activities, very little of this benefits private companies,' Green Resources stated.

> We estimate that private companies receive less than 2 percent of the public funding available for forestation and carbon. In order to increase all the activity aimed at combating climate change, in particular in Africa, funding agencies should provide much increased grants to the private sector (Carbon Positive 2010).

According to the company, project costs per hectare ranged from US$400 to US$600 (Green Resources 2008), but the land is leased for a 99-year period from the Tanzanian government at just 2.3 Norwegian krone – less than US$0.36 – per hectare, generating just under US$4.200 per year for Idete's lease (NorWatch 2009). Though this represents a decrease of two-thirds from previous land lease prices, former managing director Ivar Løvhaugen strongly criticised even this miniscule amount of money for the host country, arguing that lease prices should be reduced as much as possible to diminish risk. His thoughts were echoed by John P. Haule, managing director of a subsidiary owned by Green Resources Ltd. (then known as Tree Farms), who announced that leasing charges must decrease by 50 percent to 750 Tanzanian shillings (NorWatch 2009).

Green Resource's head, Mads Asprem, further explained to Norway's civil society group NorWatch last year:

> It is interesting to note that the demand for land in Tanzania is very low and that there is little development within forestry and agriculture. The conclusion can only be that the price of land is too high (NorWatch 2009).

The company has pledged to reinvest 90 percent of revenue back into further projects and said the two villages leasing land will receive 10 percent of the profit generated from carbon credits (Green Resources 2009). But according to Karumbidza,

> What Green Resources is doing is exporting the problem of pollution generated abroad to Africa. Tanzanians are receiving little in the process. This will become more evident in ten or 15 years when groundwater is depleted by wood plantations. The exploitative nature of the deal is especially evident in the fact that it was negotiated not in hard currency but Tanzanian shillings [which are] subject to currency depreciation. Tanzanian communities can expect to receive several million Tanzanian shillings from the carbon credit revenue in 15 years – whatever that is worth (Sharife 2010a).

Although Green Resources – the first company to receive Voluntary Carbon Standard (VCS) certification outside of the USA – has claimed that thus far the company has reaped no profits after 12 years of operation in Africa, plantations will soon be fully grown and ready to harvest (Carbon Positive 2010). Time will tell whether reforestation CDM projects are 'sustainable', but to add to the doubts, consider the deeper theory and political history of carbon trading.

The price of pollution and the privatisers of the air

REDD projects of this sort are not unusual within the industry now attempting to commodify the air. Jack Cogen, president of NatSource, the world's largest buyer of private carbon credits with more than US$1 billion in monetised 'natural' assets, bluntly characterised the mercenary nature of the business when he said: 'The carbon market doesn't care about sustainable development.' (Wysham 2005). 'All it cares about,' he continued, 'is the carbon price.'

The idea of pollution trading to solve the climate crisis was muscled into the Kyoto Protocol by the then US Vice President Al Gore, using the sulphur dioxide model of the amended Clean Air Act (1990). The basic principle is colonisation of the atmospheric commons by privatising access. This is accomplished by granting major polluters – such as multinationals exploiting fossil fuels – durable and enforceable property rights to the atmosphere under the guise of incentivising the reduction of carbon, which accounts for 75 percent of greenhouse gas emissions.

The logic of the cap and trade system – managing negative commons, or dumps – requires multinationals to purchase credits or allowances, should companies exceed government allocated caps. These credits, mainly financed by taxpayers and consumers in developed countries, are often provided free of charge to polluters. In a 2009 legislative strategy by U.S. Congress members Henry Waxman and Ed Markey, the US permits were estimated at US$646 billion in a potential worldwide market of US$3 trillion by 2020. But in 2012 alone, 69 percent of the first year permits (valued at US$55 billion) would have been freely peddled to industry, with multinationals receiving 15 percent in free permits to 'cover their increased costs from the global warming protection programme' (Johnson 2009).

Yet despite permits constituting 'deferred taxes', it is not government that quantifies the level of emissions, but multinationals themselves. This facilitates a process where corporations inflate emissions, paving the road for two possibilities: allowing multinationals to control the price of pollution, artificially depreciating it via mass sell-offs (as has been the case in Europe's stagnant markets), or alternatively, creating artificial scarcity in order to profit from stashed permits. Since permits have been commodified, cancellation or withdrawal requires governments to finance corporate losses, once again using taxpayers' funds. This of course brings back unpleasant memories of the US taxpayer-funded 'bank bail-out.' Although Goldman Sachs – which played a key role in creating the speculative bubble in the US housing market that sparked the worldwide economic downturn and the ensuing hardship for millions – is posting profits, the masses who lost their homes were not so lucky.

According to David Victor, 'the initial allocation creates expectations that the permits will be durable assets – governments might adjust allocations around the margins, but once distributed, the permits would become assets that, like other property rights, owners will fight to protect' (Victor 2001). Crucially, it also legitimises the rights of polluters to monopolise the negative commons by exhausting already decreasing atmospheric capacity (projected at 500 billion metric tons in the next four decades at present rates), preventing former colonies and emerging nations from using their share of carbon sinks. And as most governments withdraw free meal tickets from multinationals with low carbon footprints, permits actually incentivise the use of fossil fuels, undermining innovation in renewable technologies.

In his position as chairman of the International Emissions Trading Association – which represents more than 170 corporate bodies – NatSource president Jack Cogen's logic of markets resonated perfectly with corporate entities interlocked at the diplomatic- and board level with revolving door officials. Likewise, it is a logic perfectly integrated with corporatised climate discourse.

Conveniently, for example, there was little mention of the real roots of global warming (i.e. corporate exploitation) in Nobel Prize recipient Al Gore's Oscar-winning film, *An Inconvenient Truth*. Kathleen McGinty, NatSource's vice president of asset management, was the former aide to Senator Al Gore before rising to the position of key environmental advisor during Bill Clinton's presidency. Goldman Sachs holds a 10 percent share in Al Gore's Chicago Climate Exchange – the US's pilot carbon trading program, which effectively went bankrupt in late 2010 due to the lack of US climate legislation mandating cap and trade systems. Meanwhile, Goldman Sachs employees like Ken Newcombe, former head of the World Bank's Climate Change Capital, remain extremely influential.

Initially, the US's sulphur programme, the progenitor of Kyoto's cap and trade scheme, was shaped by Enron, the disgraced US energy giant that began as an obscure natural gas pipeline corporation based in Texas. The scheme enabled the corporation to dominate the newly created US$20 billion commodities market facilitated by the revamped Clean Air Act's sulphur trading programme, which authorised emission trading (US EPA 1990; see also REEP 2000).

Since the late 1970s, the US Environmental Protection Agency, mandated with the right to regulate emissions, initiated an offset scheme for power plants allowing multinational corporations to engage in modifications aimed at enhancing 'efficiency' of existing plants. The lack of 'trade' at that stage was due to the dearth of technology able to quantify emissions. 'Not much has changed despite corporate claims to quantifying tools as justifying factors,' wrote Larry Lohmann, editor of *Carbon Trading: A Critical Conversation on Climate Change, Privatization and Power.* 'No one has any idea how to calculate [the climatic value of biotic] offsets. Nor is it likely they ever will' (Lohmann 1999).

Enron's success in deregulating the trade in electricity came courtesy of intensive lobbying in Washington, by revolving door personnel like Wendy Gramm, as well as the financial wizardry of economic mercenaries stationed at Arthur Andersen, previously one of the world's major accounting firms. Five weeks prior to joining Enron's board of directors, Gramm – chairwoman of the Commodity Futures Trading Commission since the late 1980s days of George H.W. Bush's vice presidency – forced through key regulatory exemptions shielding Enron's books from public interrogation and government oversight. As early as 1989, Gramm issued policy memos lobbying for deregulation. Later, legislation pushed through Congress was delivered by Gramm's husband, Texas Senator Phil Gramm, who by 2008 had become a leading official at the Union Bank of Switzerland, Europe's largest bank. In the Senate, he authored and co-sponsored the 11 000 page Commodity Futures Modernization Act, which prohibited government from regulating derivatives and other toxic financial instruments, such as trading of energy commodities (Fox 2003).

According to emails made public following investigations on Enron, the gap in the Act (US Congress 2000:262) dumped on Congress a few days before Christmas in 2000 – the Enron 'loophole' – was backed by important agents of deregulation such as then-Treasury Secretary, Larry Summers, and US Federal Reserve Chairman, Alan Greenspan (Corn 2008). Excerpts from the email (available at http://www.enronexplorer.com) make revealing reading:

> Gramm needs to fully understand how helpful the bill is to Enron … The legislation eliminates concerns that our derivatives transactions may be illegal or that our online platforms may be unregulated futures exchanges … Without this legislation the industry will be crippled … Senator Gramm has been contacted by a number of people urging passage of the bill. This clearly has had an impact. Senator Gramm appears committed, and he and his staff are very focused … As the Bush transition moves forward, we need to monitor (or perhaps try to influence) the selection of the new CFTC Chair/Commissioners (Lipton 2008).

Arthur Andersen's and Enron's fall from grace signify deviant acts on the part of corporate mercenaries, yet such activities remain endemic to the global financial architecture. The International Accounting Standards Board, founded and financed by the 'big four', deliberately prevents corporate country-by-country reporting, which would indicate economic activities in host and home countries. This practice enables corporations, which now conduct 60 percent of global trade within – rather than between – entities (Sharife 2011), to self-regulate global trade. Unsurprisingly, the big four – in addition to 430 major banks, 70 000 corporations, 720 insurance firms, among other trusts, funds, foundations and corporate vehicles – are registered in the world's largest secrecy jurisdiction, the Cayman Islands (*Ecay Business Online* 2011). Under such an arrangement, little will trickle down to the Third World.

Gaming carbon trading

It was at the Rio Earth Summit, where the UNFCCC was adopted, that officials of the Chicago Climate Exchange (CCX), self-described as 'the world's first, and North America's only, legally-binding integrated emissions-reduction-, registry and trading system', delivered a paper advocating 'a market-based solution to global warming' (Knoll 2007). Founded just prior to the signing of the Kyoto Protocol by Richard Sandor, an economics professor known as 'Mr. Derivatives', the CCX owns the European Climate Exchange, is intimately linked to Gore, and had ties to the upper echelons of the UN. The founding members include major companies, such as Ford, DuPont, and Goldman Sachs, all of which were instrumental as leaders in designing and financing the carbon market system (Sharife 2010c).

The collapse of the CCX in October 2010, when it announced it would no longer be trading carbon, was viewed as a huge setback for both the US and the global industry. American President Barack Obama was widely reputed to have been a member of the Joyce Foundation, which funded CCX and Sandor with grants. Surprisingly, no mainstream US media outlets appeared to report the whys and hows of the collapse, with the issue barely rendering a mention despite climate policy and business frequenting the news. Initially, CCXs projected profit margins ranged in the hundreds of billions.

When Sandor – listed by *Time* magazine as one green-wrapped 'hero' of the world for founding CCX – sold his 16.5 percent stake, he netted US$98.5 million (Ellison 2002). 'The power of the free market is that it can restore nature's wealth as it increases financial wealth,' he said then. Crain's *Chicago Business* Washington, D.C. bureau chief Paul Merrion (2010), comforted readers with the statement that CCX would 'continue trading carbon offsets generated by projects that consume greenhouse gases, such as planting trees'.

Much like the derivatives bubble that caused the global recession, the carbon market and Clean Development Mechanism projects are vulnerable to gaming. One paper, 'Scaling The Policy Response To Climate Change', by Benjamin Sovacool and Marilyn Brown (2009), revealed a €4.7 billion scam structured around trifluoromethane (also known as fluoroform) (HFC-23), a greenhouse gas used as a refrigerant. More than 70 percent of CDM projects were based on HFC-23, a chemical that was deliberately produced in excess by corporations that then claimed to offset it in order to receive financial benefits via certified emissions reductions (CER) certificates. Sovacool and Brown's study also evaluated 93 randomly selected CDM projects and found that 'in a majority of cases, the consultants hired to validate CERs did not possess the requisite knowledge needed to approve projects, were overworked, did not follow instructions, and spent only a few hours evaluating each case' (Sovacool and Brown 2009).

The UNFCCC has registered 2 582 projects and issued 476 762 324 CERs (UNFCCC 2010). The carbon offsetting organisation Carbon Retirement revealed that 28 percent of total funds received for CDM projects went to developing countries; 30 percent was received by banks and investors; 17 percent to company shareholders; and the remainder to taxes. The World Bank has proposed, through the Climate Investment Funds unit, that the Bank levy a fee of US$350 000 for each investment project, 40 percent more than standard development project fees (Bretton Woods Project 2010). The Bank's president, Robert Zoellick, has promoted the Bank to developing countries as the ideal vehicle to engage in climate financing, a strategy laid out in leaked World Bank documents (Sharife 2010c).

But it is not only through the World Bank that the foreign policies of developed nations are implemented. Aid represents another key vehicle. As *WikiLeaks* recently exposed, US Under secretary of State for Democracy and Global Affairs, Maria Otero, urged the Ethiopian premier and head of the Africa Union climate change negotiation's team, Meles Zenawi, to sign the Copenhagen accord, as it represented a point of departure for other discussions – or else face potential cuts in US aid funds (*The Guardian* 2010).

Even worse, the financing that developed countries persistently claim cannot be raised, could in fact be generated easily through a small financial transaction tax, either on cross-border financial flows – the Tobin Tax – or as has recently been developed in the UK on a wider range of domestic transactions – the Robin Hood Tax. Campaigners say such a tax could generate US$400 billion, chiefly from developed countries. Not only would this require information sharing on a multilateral basis as implemented within international agreements, but it would also allow for developing and developed nations to track and recover stolen wealth (Sharife 2010c).

The issue of climate must be broadened beyond 'adaptation and mitigation' to a systemic overhaul, beginning at the fault line: the lack of political will on the part of systemically powerful polluting countries and corporations to correct imbalances of power in the market, and the unregulated nature of international trade and financial transactions. In this context, the UN's acceptance of cap and trade is debilitating. As far back as the UN's Rio summit in 1992, Gore's CCX declared the need to realise a 'market-based solution to global warming' (*Reuters* 2008).

This ideology bears resemblance to the propagation of the common cuckoo: brooding parasites that deposit eggs in the nests of other birds. Carbon capitalism destroys and marginalises other living beings, voraciously feeding off host entities at the expense of host communities. If scientists are correct and the atmosphere can withstand only 250 billion metric tons more in our carbon budget before the planetary ecosystem hits bankruptcy, our choices, until the earth recovers, are limited to a zero-carbon future – beginning with drastic cuts in developed, not developing, countries. This means doing away with carbon capitalism and capitalism overall. The only other alternative is a zero-future.

References

Ababa, A. 2007. Ethiopia loses 200,000 hectares forest annually. *The Ethiopian Herald*. February 8.

Bond, P. 2007. Privatization of the air turns lethal: "Pay to pollute" principle kills South African activist Sajida Khan. *Capitalism Nature Socialism* 18 (4): 6–37.

Bond, P. and K. Sharife. 2011. At the heart of our wealth and our woes, the "Minerals-Energy Complex." *The Mercury*. July 19.

Brown, S. 1999. Guidelines for inventorying and monitoring carbon offsets in forest-based projects. Report prepared for the World Bank. Winrock International.

Business News Network. 2010. Interview with Green Resources' Mutuma Marangu. April 28, http://tinyurl.com/2f34wg9.

Bretton Woods Project. 2010. Update on the Climate Investment Funds. July Summary.

Carbon Positive. 2010. Reforestation vital for Africa & climate developer, http://tinyurl.com/y9qhuvf.

Christian Aid. 2009. The role of carbon markets in countering climate change. Signposts to Copenhagen 3. August, citing M. Wara and D. Victor. 2008. A realistic policy on international carbon offsets. Program on Energy and Sustainable Development Working Paper 74, Stanford University. April.

Corn, D. 2008. Foreclosure Phil. *Mother Jones*. July/August.

Daily Nation. 2009. Lush and expansive tea plantations conceal damage to forest. August 2, http://www.nation.co.ke/News/-/1056/633794/-/uljv0b/-/index.html.

ECay Business Online. 2011. Offshore financial services in the Cayman Islands.

Ellison, K. 2002. Richard Sandor: His market is a gas. *Time*. August.

Eraker, H. 2000. CO_2lonialism: Norwegian tree plantations, carbon credits and land conflicts in Uganda. NorWatch. April: 4, *www.norwatch.no/...in...treefarms-in-uganda-co2lonialism/download.html.*

Fox, L. 2003. *Enron: The rise and fall*. New York: John Wiley and Sons.

Global Witness. 2008. Honest engagement: The need for transparency and civil society participation in REDD. December.

Green Resources. 2008. Overview of plantation/certification development in Tanzania: 13–14.
———. 2009. CCA Africa agribusiness sector presentation.
———. 2010. Plantations: Tanzania, http://www.greenresources.no/Plantations.aspx.
———. 2011. Major shareholders, http://www.greenresources.no/Company/Shareholders.aspx.

Greenpeace. 2003. Illegal logging and the heart of darkness. April 22, http://www.greenpeace.org/international/en/news/features/illegal-logging-in-cameroon/.

Johnson, K. 2009. Waxman-Markey bill: It's giveaway time for emissions permits. *Wall Street Journal.* May 15.

Knoll. 2007. Former President Clinton Recognizes Knoll and the Chicago Climate Exchange at Greenbuild. November 7, http://www.knoll.com/news/hstory.jsp?story_id=4818&type=Press%20Releases&storyType=nf.

Lang, C. 2004. The carbon spin doctors: How the World Bank explains emissions trading to journalists. *World Rainforest Movement* 84, http://www.wrm.org.uy/bulletin/84/carbon.html.
———. 2009. Ogiek threatened with eviction from Mau Forest, Kenya. *redd-monitor.org*. November 19, http://www.redd-monitor.org/2009/11/19/ogiek-threatened-with-eviction-from-mau-forest-kenya/.

Lipton, E. 2008. Gramm and the Enron loophole. *The New York Times*. November 17.

Lohmann, L. 2006. Carbon trading: A critical conversation on climate change, privatization and power. *Development Dialogue* 48. September, http://www.free-books.us/Others/31919/Carbon-Trading-A-Critical-Conversation-on-Climate-Change-Privatization-and-Power.

Macken, K. 2011. Strengthening credibility in the E.U. ETS following security and fraud related incidents. *International Network of Environmental Compliance and Enforcement*, http://inece.org/conference/9/papers/Macken_Ireland_Final.pdf.

Merrion, P. 2010. Chicago Climate Exchange pares more jobs. *Chicago Business*. November 2, http://www.chicagobusiness.com/article/20101102/NEWS01/101109966/chicago-climate-exchange-pares-more-jobs.

Ndungu Commission of Inquiry. 2004. Report of the Commission of Inquiry into the illegal/irregular allocation of public land, http://marsgroupkenya.org/pages/stories/Ndungu_Report/index.php.

NorWatch. 2009. Climate project on cheap ground. June 5, http://www.norwatch.no/200906051306/english/other/climate-project-on-cheap-ground.html.

Oliver, R. 2008. Rich, poor and climate change. *CNN*. February 18, http://edition.cnn.com/2008/BUSINESS/02/17/eco.class/.

Ostrom, E. 2009. A polycentric approach for coping with climate change. World Bank Policy Research working paper. October.

Reddy, T. 2005. Durban's perfume rods, plastic covers and sweet-smelling toxic dump. *Carbon Trade Watch*. March 9, http://www.carbontradewatch.org/articles/durbans-perfume-rods-plastic-covers-and-2.html.

Renewable Energy Policy Project. 2000. A guide to the Clean Air Act for the renewable energy community: Emissions trading under the Clean Air Act, http://www.repp.org/repp_pubs/articles/issuebr15/03emTrad.htm.

Republic of Tanzania. 2007. A handbook for Clean Development Mechanism projects (CDM) activities in Tanzania, http://tinyurl.com/27bo4kd.

Reuters. 2008. Climate exchange PLC – Preliminary results. March. 12, http://uk.reuters.com/article/2008/03/12/idUS69488+12-Mar-2008+RNS20080312.

Rift Valley Holdings Ltd. 2011. About Us, http://www.riftvalley.com/.

Science Daily. 2009. One-fifth of fossil-fuel emissions absorbed by threatened forests. February 19.

Sharife, K. 2010a. Author interview with Dr. Blessing Karumbidza. October 2.
———. 2010b. Pulp fact or fiction? Clean Development Mechanism (CDM) plantation projects. *Pambazuka News* 499. October 7, http://www.pambazuka.org/en/category/features/67534.
———. 2010c. Unpacking the hot air industry. *The Africa Report*. December.
———. 2011. "Transparency" hides Zambia's lost billions. *Pambazuka News* 536. June 23, http://www.pambazuka.org/en/category/features/74275.
———. 2009. 'What is the real value of Africa's wealth". *African Business*. May.

Sovacool, B.K. and M.A. Brown. 2009. Scaling the policy response to climate change. *Policy and Society* 27: 317-328.

The Guardian. 2010. U.S. embassy cables: U.S. urges Ethiopia to back Copenhagen climate accord. December 3.

Tetchiada, S. 2005. Environment-Cameroon: Corruption rooted in logging industry – NGOs. *Inter Press Service*. July 18, http://ipsnews.net/africa/nota.asp?idnews=29537.

TimberWatch. 2009. "Potential impacts of tree plantation projects under the CDM: an African case study. http://tinyurl.com/2dsqo3h.

UNEP. 2006. COP12 UNFCCC/MOP2 Kyoto Protocol: Challenges and opportunities for Africa. *ROA News*. December 10, http://www.unep.org/roa/docs/ROA_news/ROANewsNo_10.pdf.

UNEPFI. 2010. Investors representing $13 trillion call on U.S. and other countries to move quickly to adopt strong climate change policies. January 14, http://www.unepfi.org/fileadmin/documents/14jan_investor_statement_pressrelease.pdf.

UNFCCC. 2010. CDM registry, http://cdm.unfccc.int/Registry/index.html.

U.S. Congress (106th). 2000. H.R. 5660: Commodity Futures Modernization Act of 2000. Library of Congress, http://thomas.loc.gov/cgi-bin/bdquery/z?d106:H.R.5660.

U.S. EPA. 1990. The Clean Air Act Amendments of 1990, http://www.epa.gov/air/caa/overview.txt.

Victor, D. 2001. *The collapse of the Kyoto protocol and the struggle to slow global warming*. Princeton: Princeton University Press.

Vidal, J. 2009. Redd in Africa: How we can earn money from air by harvesting carbon. *The Guardian*. October 5, http://www.guardian.co.uk/environment/2009/oct/05/redd-kenya-climate-change.

Wafula, C. 2009. Family given nearly all of Maasai forest. *The Nation* (Kenya). August 4, http://allafrica.com/stories/200908050153.html.

Wysham, D. 2005. Carbon trading: A planetary gamble. *Fort Worth Star-Telegram*. December 7.

Chapter 8

The Endless Algebra of Climate Markets

Larry Lohmann[1]

When Sir Nicholas Stern, climate change advisor to Tony Blair's government, famously said in 2007 that global warming was 'the greatest market failure the world has ever seen', the implication was that, given the proper price signals, addressing it could be a great market success. With appropriate information, property rights, regulation and taxes – along with new markets for greenhouse gas pollution – the task of prying apart fossil fuels from economic life and thus *decarbonising* society could be largely internalised into everyday market activity (Stern 2009:11). Predictably, Stern, a former World Bank economist and vice-president, became a climate businessman himself, and today, in addition to his other jobs, serves as an advisor to IDEAcarbon, a company whose ambition is to provide 'ratings, research, and strategic advice' on carbon commodities and finance to 'buyers, sellers, and hedgers' (IDEAcarbon 2011).

Who built the US$hundred billion carbon markets that Stern and so many others have heralded, and why? International climate trading – one of many neoliberal environmental innovations of recent times – began to be planned and developed in earnest not long after the institution of the US sulphur dioxide market in the 1990s. Background conditions included – in addition to the ever-growing influence of neoliberalism – increasing financialisation, the imperative for surplus capital to find new outlets at a time of declining profits, and the continuing global political dominance of the USA.

But, as is always the case, there were also more specific and idiosyncratic drivers; and those drivers, and the materials they interact with, have changed over time. Carbon trading has had not one progenitor, but many; and not one history, but many. Insular (even clubby) as the expert elite responsible for the first stages of its development is, it has always straddled a number of conventional divides, drawing participation from

— among others — the financial sector, NGOs, think tanks, university economics departments, government lawmakers, certain business sectors, law and consultancy firms, and the United Nations system.

A few examples will suggest the extent of this diversity. Richard Sandor, a wealthy Chicago trader and economist who was one of the originators of interest rate derivatives in the 1970s, helped elaborate the theory of pollution markets in the 1980s, collaborated on a UN Conference on Trade and Development (UNCTAD) initiative entitled 'Building a Global CO_2 Emissions Trading System' in the early 1990s, and, in the 2000s – with philanthropic support – set up the Chicago Climate Exchange, of which he sold off his shares in 2010 for US$606 million (Carr and Lomax 2010). At UNCTAD, Sandor worked under Frank Joshua, who later became the global director for emissions trading services at Arthur Andersen before joining NatSource, a big carbon commodity trader. Also at UNCTAD was Alice LeBlanc, at the time a staff member of the corporate-friendly Washington NGO Environmental Defense — an organisation that had helped write the legislation for the US's sulphur dioxide market in the 1990s. LeBlanc later joined Sandor at the Chicago Climate Exchange before becoming head of the climate change office of the ill-fated insurance- and speculation firm AIG.

Figures such as Robert Stavins, a neoclassical economist at Harvard who had also played a part in the US sulphur dioxide programme, meanwhile provided support from academia, where the whole idea of pollution trading had been hatched in the 1960s. Across the Atlantic, Michael Grubb of the Royal Institute for International Affairs, proposed that the sulphur dioxide scheme could serve as a model for a world carbon market.

Ted Hanisch, a Norwegian government official, explored ways in which carbon trading could provide a way for his country to 'compensate' for its oil production and industrial and shipping emissions. In the mid-1990s, Graciela Chichilnisky, a high-powered mathematician and neoclassical economist at Columbia University, pitched the idea to US and UN officials, claiming the mechanism could redistribute wealth to the global South (Chichilnisky 2009).

In 1997, the Clinton regime, represented by Al Gore (who later went into the carbon business himself as a private individual), played the decisive role in ensuring that the Kyoto Protocol became a plan for a world climate market. Although the George W. Bush regime withdrew from the Kyoto agreement in 2001, much to the dismay of some US trading firms such as Enron, market development continued under figures such as Ken Newcombe, who headed the World Bank's Prototype Carbon Fund before moving on to Climate Change Capital (a boutique merchant bank founded by, among others, attorney James Cameron, who had also helped negotiate the Kyoto Protocol as an advocate of the interests of small island states), Goldman Sachs' carbon trading desk, and the carbon trading firm C-Quest Capital (C-Quest Capital 2011).

Such cross-sectoral lines of influence continue to be crucial to climate market construction at all levels. For example, the current Executive Secretary of the UN Framework Convention on Climate Change, Christiana Figueres, was, up until the time of her appointment in 2010, senior advisor to C-Quest Capital, principal climate change advisor to Endesa Latinoamérica, the largest private utility in Latin America, and the vice-chair of the rating committee of the Carbon Rating Agency, a private firm associated with Lord Stern's IDEAcarbon (Figueres 2011). Earlier, Robert Stavins's students and colleagues had influenced carbon market development in the EU, which, unable to get agreement on a European carbon tax, adopted the EU Emissions Trading Scheme (EU ETS) as the centrepiece of its climate policy in the 2000s, engendering what is today the world's biggest carbon market, accounting for 97 percent of world market volume (World Bank 2011).

What are carbon markets for?

> No bell rings to mark the end of building and the beginning of functioning (Daniel C. Dennett 1995:218).

Reflecting the divergent interests and *bricolaged* evolution suggested by such vignettes, the rationales offered for carbon markets tend to be varied, slippery, and shifting. They do have a common basis: an abstract, idealised distinction between market and non-market, between a unitary 'capitalism' and what is 'outside capitalism', and between 'externalisation' and 'internalisation'. But the contested, incomplete, and ultimately incoherent nature of these divides precludes them from providing a basis for constructing a firm or lasting justification for the new trade. Over time, it becomes obvious that there is no single, settled explanation of what carbon markets are *for*.

One relatively lasting claim, developed in an earlier phase of the history of emissions trading during the 1970s and 1980s, has been that carbon markets are technical instruments for 'saving costs' in achieving an exogenous goal of 'climate change mitigation'. It is admitted, of course, that setting up carbon markets in the first place is far from cheap. Governments have to spend money setting and enforcing progressively stricter overall sectoral or societal caps on emissions. They also have to take on the task of dividing up emissions quotas among the industries under their jurisdiction, and setting up the legal and quantificational machinery for making them tradeable (Cole 2002). But, the argument goes, trading then softens the blow for the private sector, making the achievement of reductions more efficient and allocating the earth's carbon-cycling capacity to where it will be most productive. Industries for which reductions are expensive can save money by buying extra pollution rights from industries for which they are cheap; those industries in turn can make money by selling permits that they don't need.

This justification has come to be supplemented by a good half dozen more. First, science-based policy: under emissions trading, government mandated emissions reductions informed by climatology determine the carbon price rather than a government mandated carbon price (via, say, a carbon tax) determining emissions reductions. This is said to be a more direct way of achieving the needed reductions than having to adjust and readjust prices constantly to see which one might yield the scientifically 'correct' cuts.

Second, innovation: although buyers of emissions rights can evade some of the burden of low carbon innovation, prospective sellers seeking economic advantage eagerly take it on. Even if it cannot wean industry off fossil fuels by itself, it is said that 'giving carbon a price' will thus help drive the needed energy revolution.

Third, information: instead of an information challenged state having to 'choose technology winners', an information rich, decentralised 'market' is said to do so instead, resulting in added efficiencies as well as transparency.

Fourth, profit: 'internalisation' of the global warming 'externality' makes new forms of accumulation possible, and in the process helps to create a revamped, sustainable 'capitalism 2.0' (or 3.0, 4.0 or 5.0).

Fifth, deregulation: with carbon markets in place, governments can avoid or roll back allegedly clumsier 'command and control' types of regulation, saving money for themselves as well as for business at a time of profit crisis. They can also avoid having to subsidise climate action insofar as much of the necessary funding and revenue will be provided by the carbon markets.

Sixth, political convenience: by making climate action cheaper or even profitable, carbon trading supposedly makes it more palatable for industry, government, and the public. Indeed, once the new carbon assets are in circulation, even the powerful financial sector should throw its political weight behind radical action on global warming, since it is to its advantage to speculate and act as an intermediary in the new trades (Newell and Paterson 2010). Capitalists, it is said, will rally behind carbon trading, since they always favour market approaches over others.

This classic neoliberal script, which has been slowly constructed over decades, runs into problems when the materials and *dramatis personae* out of which it is fashioned find themselves enacting a series of contradictory narratives as well, while additional actors, all with their own agendas, also join the show. For example, once carbon pollution rights become an asset, a stubbornly fossil fuel dependent industrial sector becomes as enthusiastic about securing maximum free allocations from governments, as it ever was in urging lower taxes or curbs on conventional regulation. The distinction between the design of a market mechanism and the profit seeking activities that it enables quickly goes by the board. With it goes any pretence of governments being able to set emissions caps according to scientific criteria. Influenced by the need to make concessions to owners of fossil-fuelled infrastructure as well as a host of other factors, carbon prices oscillate around a derisory level, powerless to incentivise a transition away from fossil fuel use.

Indeed, many fossil-fuelled industries use the generous handouts of pollution rights that they have been given to entrench carbon intensive 'business as usual' even more firmly. Market defenders then divide. Idealists dig their theoretical heels in, decrying the rent seeking activities of polluting companies as illegitimately 'non market' while urging governments on to heroic feats of cap setting that they show little sign of wanting to perform (e.g., Sandbag 2011). Cynics meanwhile shift their ground, falling back to the position that government's role is to provide just enough scarcity to keep a market going, while making long range provisions for carbon price ceilings to forestall disruption among industries with huge sunk costs in fossil-fuelled plants and few alternatives. 'Well, at least carbon has a price', market advocates shrug – although when prices of what has been called the world's 'worst performing commodity' (Wynn and Chestney 2011) threaten to dip to zero, even that assertion can come into question. What with all the political capital invested, the EU Emissions Trading Scheme moves 'from being a means to a carbon end to being more of an end in itself' (Helm 2010:189).

Still more slippage ensues when even the weak emissions caps to be imposed on industrialised nations prove too much to bear for fossil-fuelled industry. Search parties are sent out for cheap substitutes for emissions cuts. Enter carbon offsets, which allow Northern industries and nations to evade caps by buying pollution rights from projects outside the caps that are certified to result in less greenhouse gas emissions than would otherwise have been the case. Soon more ingenuity will be poured into offset manufacture abroad than ever went into low-carbon innovation at home. Cue the reassurance that offsets are merely a stopgap while industry finds its feet in the new regulatory environment and prepares for the steep cuts of the future. Then, as it becomes clear that offsets are, in fact, central to the market, there comes a renewed insistence that these 'substitute' emissions reductions are just as good as the real thing, only cheaper.

When the climatic ineffectiveness of offsets is subsequently exposed through a series of scandals and accounting controversies (see, e.g., *CDM & JI Monitor* 2011; Parekh 2011), yet another rationale is wheeled out: that offsets have a redistributive or 'Third World development' purpose rather than a climatic one (see, e.g., Gronewold 2011). Finally, when disgruntled host communities call attention to the fact that they do not fulfil this purpose either – since offset finance is being captured by corporate bad hats and large landowners rather than ordinary people – market defenders unveil programmes for 'participatory reforms' (e.g., UNFCCC 2011).

These are not the only ways in which the evolving carbon market organism constantly slips the surly bonds of its early, clichéd neoliberal portrayals. In yet another twist, no sooner are various carbon commodities mooted than they attract the excited attention of Wall Street and the City of London, who target them for investment, securitisation, hedging, bundling, and speculation. Soon banks and shadow banks, rather than polluting industries, are the biggest buyers of offsets. In a post-1970s era of profit crisis and financialisation, networks such as the International Emissions Trading Association (IETA) understandably find themselves more preoccupied with lobbying for a large, liquid market with plenty of opportunities for intermediation and speculation than with the climate effects of the trading system. Not only prices, but the structure of the commodities themselves, fall under the influence of derivatives traders. The early vision of a limited, technical trading device aimed at saving costs in achieving a sharply delineated emissions goal for a small class of big emitters is gradually overshadowed by that of a giant new engine for capital accumulation. The idea that climate action is being 'internalised' into market incentives becomes curiously passé. Queried about the climate effectiveness of carbon offsets, Richard Sandor retorts impatiently that whether or not they result in less greenhouse gas going into the atmosphere is 'not my business. I'm running a for-profit company' (Ball 2008). Even the penultimate default environmental justification for carbon markets – that they are 'better than nothing' – loses credibility and begins to fall into disuse. The Thatcheresque slogan that 'carbon markets may be bad for the climate, but there is no alternative' rises in its place.

Also meeting an untidy end is the narrative of a lean, clever, muscular market lifting the climate action burden from the shoulders of a clumsy, wheezing, out-of-shape, overbearing state. Carbon market requirements for state measurement, calculation, monitoring, enforcement, certification, registration, regulation, and creation of property rights (not to mention the constant diplomacy required to keep up confidence that international markets are on track) turn out to be enormous. Traders and financiers demanding maximum production and maximum standardisation of carbon commodities – and the maximum demand required to absorb them – find themselves appealing to the same state or international agency regulators as more cautious market advocates, who, concerned about quality control, stress that increased government oversight may be necessary to convince punters that the new commodities are credible (see, e.g., IETA 2010; CAN 2010). In the US, Tea Party activists soon detect unseemly similarities between carbon markets and the dread 'big government' approaches of carbon taxes and conventional regulation, and go to work to help shoot down federal plans for cap and trade. Respected voices in economics, law, and finance fret that carbon markets are even less transparent than conventional state mechanisms: 'the opportunities for graft, bribery and corruption' in the carbon offset industry are 'limitless' (Buiter 2007); that an 'opaque set of variable standards' creates a 'tremendous incentive to create (or destroy) supply as it suits the participants in the market' (The Munden Group 2011:17); that rather than clarifying decision-making, carbon trading 'provides a layer of additional complications and occasions for dispute' (Driesen 2003:94).

Their neoliberal faith unsettled, states start contemplating more legislation that subsidises renewable energy – which, naturally, turns out to threaten to undermine carbon prices even further and so ironically comes under fire (Nicholls 2011:4). Meanwhile, the new markets' ability to fund global warming action turns out to be disappointing: most market growth statistics reflect trading by intermediaries, not climate change mitigation, and the bulk of the cash laid out for supposedly carbon-saving projects goes to consultancies planning and certifying projects of dubious climate relevance. On the academic front, the whole idea of an 'economisable' climate goal meanwhile comes into question, given the uncertainties, scale and complexity of global warming (Weitzman 2008; Hulme 2009). In 2010, what with the financial crisis, a spate of criminal and other scandals, uncertainty about the future of UN climate treaties, an increasingly embarrassing lack of climate results (Brinkley and Less 2010), and growing opposition from environmental justice movements, market growth stalls in spite of active trading in 'paper carbon' on the financial markets. In the wake of the failure of the US Congress to pass legislation mandating a nationwide carbon market, even Richard Sandor sells out his shares in the Chicago Climate Exchange (Carr and Lomax 2010), while in the usually enthusiastic precincts of the City of London and

Wall Street, a new pessimism takes hold. Banks shed staff from their carbon trading desks, and some of the hundred-odd specialised carbon funds that sprang up in the 2000s close their doors (Sills 2011; McGarrity 2011). The IETA loses members, and although attempts to lay the infrastructure for commodifying the carbon absorptive capacity of global forests continue, few observers are optimistic about the outcome of future negotiations over the UN carbon market.

The final insult to neoliberal nostrums is that much of the business community itself – particularly the parts with less experience in commodity or financial trading – turns out to be unexpectedly resistant to the new carbon markets. The essentialist neoliberal claim that since carbon trading is a 'market solution', it must be in line with 'capitalism', and that therefore capitalists ought to like it, turns out not necessarily to cut much ice with corporate actors who have always done their best to avoid market mechanisms that fail to serve their purposes, while availing themselves of as many 'non market' mechanisms (violence, plunder, subsidies, corruption, enclosure of commons, patriarchy, intra-corporate coordination, etc.) as are needed to turn a buck in any particular historical conjuncture (Harvey 2010; Mitchell 2002).

Expert voices are heard cautioning against an undiscriminating enthusiasm for 'market-based instruments', noting that it may not be 'relevant and useful' to gather them 'under a single label', much less to assume that they necessarily enhance cost-efficiency, reveal information, distribute the right incentives, or correct market failures (Broughton and Pinard 2011:3–4). Despite these developments, a few critics of neoliberal climate policy, particularly in the industrial north, find themselves going along with the essentialist neoliberal claim that carbon markets are a natural, almost inevitable outgrowth of an abstract capitalism. Concluding that it is this latter elusive chimera that must be the proper target of political mobilisation, they are often drawn into an irrelevant, gaseous policy debate conducted by neoliberals eager to pit proponents of 'necessary reform' against those of 'impossible revolution'. Largely indifferent to such amusements, carbon market actors get on as best they can with the more serious business of accumulation through climate commodity exchange, with all its deleterious consequences for global warming and human welfare.

The foregoing pastiched history may appear to depict nothing more than an unusually extreme version of a standard neoliberal comedy of unintended consequences. However, the very scale of the failure of carbon markets to do what they were first advertised to do suggests that their history will provide especially clear-cut tools for problematising the metaphors that accompany the booming trend of market environmentalism as a whole. And so it proves. In addition to exposing in particularly stark relief how the contradictions of attempts to 'internalise externalities' and institute 'correct environmental pricing' can unfold, an examination of carbon market dynamics can also clarify the misfortunes that befall the notion that there exists a fixed hardware of 'an economy' that can be reprogrammed with the green software of 'natural capitalism' in a way that 'fully incorporates nature into its system of value' (Paul Hawken, quoted in Foster 2003), or the notion that a steady state economy or (alternatively) a regime of green growth could someday bring the history of environmental struggle to a happy end.

Making a market from climate crisis: A thought experiment

How do you make a market out of climate? The answer is not immediately obvious. Global warming results mainly from the transfer of carbon from a fossil pool locked underground to a separate pool circulating above the ground among the atmosphere, oceans, vegetation, soils, fresh water, and surface rocks. This transfer is irreversible over humanly relevant time scales. It follows that sustaining – or 'producing' – the use-value of a liveable climate requires keeping remaining fossil fuels in the ground.

To put it another way, given path dependence (Arthur 1994) and the way that fossil fuels have become 'locked in' (Unruh 2000:817) to industrialised societies' ways of life, it calls for political mobilisation behind immediate long-term investment programmes in new, non-fossil energy, transport, agricultural, and consumption regimes, particularly in the North, as well as programmes for shifting state subsidies from fossil fuels to existing initiatives defending or constructing low carbon means of livelihood. Above all, it demands widespread alliance building in support of the social movements that are already directly or indirectly addressing the below- to above-ground transfer of carbon. These include movements working to 'keep the oil in the soil, the coal in the hole, and the tar sand in the land' in the Niger Delta, Alberta, Ecuador, South Africa, Appalachia, and elsewhere; stopping the development of dozens of coal-fired power plants in the US, Britain, Thailand, and other countries; fighting agrofuel projects whose effect would be to sustain a transportation infrastructure designed for oil; and working to ban banks from supporting fossil intensive or fossil extractive projects. Increasingly, such movements are aligning themselves with those in support of ecological and peasant agriculture, more democratic public health and energy provision, cleaner air and water, and an end to militarism, environmental racism, and extractivism.

Prima facie, the carbon market's objective should be to support movement building of this radical kind. Yet how might it be possible to buy and sell contributions toward the long-term political shift away from fossil fuels that such movements are working toward? Yale law professor Douglas Kysar suggests that the 'legal and political actions' that have 'dramatic impact' on historical trends would have to be commodified. The resulting climate commodities could be sold by, for example, 'indigenous groups that entirely block new exploration activities' or 'forest dwelling communities that successfully fight to stop logging' (Kysar 2010). In this delirious vision, Lloyd Blankfein and his colleagues at Goldman Sachs, sitting down to craft products for sale to investors, would 'devote themselves … to the identification and promotion of critical political interventions by disempowered voices for sustainability' (Kysar 2010). Accumulation would be a matter of investing in commodities that maximised radical and structural societal change over the long term.

Kysar's tongue-in-cheek thought experiment hints at the novel and extreme ways in which the contradiction between use-value and exchange-value, between the qualitative logic of ensuring survival goods and the quantitative logic of profit, will unfold and extend itself in climate markets. To make accounting, ownership, and capital accumulation possible, Kysar's climate market would have to turn the qualitative relations that make up movement building and historical process into quantitative ones. For example, consumers would need to know, and producers to guarantee, what increment of historical change toward a halt to fossil fuel extraction each commodity sale represented.

But who would quantify the extent to which each unit of the commodity contributed to undoing the social complexities of fossil fuel path dependence, and how? If different units contributed different increments of historical change depending on the particular pathway they were aggregated into, and the paths were incompatible, how would the units be commensurated? How would the historical effects of private ownership on the dialogue and movement building comprising the 'labour' producing the climate commodity be calculated? (For example, would street demonstrators wearing corporate logos on their T-shirts lose their effectiveness?) If the experts in counterfactual history that were called upon to help price the commodity attached a particular value to rolling back the dominance of a rampant financial sector, would Goldman Sachs still sell the associated securities? And so on.

The only way of removing such difficulties for accumulation would be to demote the market to being a provider of unspecified and unquantifiable 'climate services' – in which case it would lose most of its usefulness for policy makers and its appeal to other potential customers.

A more 'realistic' model

> Those combinations of theory and empiricism that conform to the arbitrary definitions of the commodity will clearly be preferred to those that stress complexity and the interrelatedness of phenomena (Philip Mirowski 2011:206).

The alternative to the immediate, dizzying multiplication of paradoxes of Kysar's whimsical proposal is to construct a market based on the enclosure and commodification of pollution sinks, whose extent the state defines in terms of limits on the quantity of molecules that can be emitted. This is what the US's sulphur dioxide trading system instituted in the 1990s did, and it is the model followed by the Kyoto Protocol's carbon market, the EU Emissions Trading Scheme, and all other actually-existing climate markets.

The advantages are obvious. Molecules can be counted (in many pollution markets, a metric ton is the unit of measurement). Molecules come 'pre-standardised' in the sense that they are the same the world over. Molecules can also be owned. So, at least in principle, can the sinks that absorb them – for example, oceans, trees, or land that absorb carbon dioxide. Quantifiability and ownability make it possible to buy and sell rights to emit CO_2 – essentially, rights or access to the earth's carbon cycling capacity in the oceans, the atmosphere, soil, vegetation, and rock. And with measurement and private property come, too, the possibility of large-scale accumulation. This is why actually-existing climate services markets, instead of being structured in a way that would support existing grassroots movements, are based on the equations:

$$\text{a better climate} = \text{a reduction in } CO_2 \text{ emissions}$$

and

$$CO_2 \text{ reduction A} = CO_2 \text{ reduction B}$$

But if there is to be a market in CO_2 emissions reductions, someone must 'produce' them, and someone must buy them. To put it another way, if there is to be a market in greenhouse gas pollution dumps, someone must make them scarce, someone must 'own' them, and someone must 'rent' them. Setting up this apparatus can only be the job of governments, who must impose both the need for reductions (by making pollution dumps scarce) and the means of 'producing' or owning them. Governments achieve the former by imposing 'caps' or limits on emissions on companies or economic sectors. To accomplish the latter (that is, create a reduction commodity), governments need the additional equation:

forced reduction of CO_2 emissions to level c within time period p	=	tradeable right to emit CO_2 up to level c by the end of period p

Carbon dioxide reductions (and by inference, climate action) can accordingly be achieved by the production of tradeable pollution rights, whose scarcity or otherwise is determined by government fiat. *Progressive* carbon dioxide reductions can in turn be achieved by relying on the equation:

reducing CO_2 emissions progressively	=	issuing fewer tradeable rights to emit CO_2 in period $p + 1$ than were issued in period p

The producers or owners of these rights are, in the first instance, governments themselves. European Union Allowances, for example, are 'produced' in a preset amount by the pens or keystrokes of politicians and bureaucrats under the European Union Emissions Trading Scheme (EU ETS). They are then sold or, more usually, given away free to large private sector polluters (a dramatic rise in rent charged for natural resources, after all, has always been one of the most feared barriers to capital accumulation). Assigned Amount Units (AAUs), one of the climate commodities of the Kyoto Protocol carbon market, are meanwhile 'produced' by conferences of the parties to the UN Framework Convention on Climate Change before being distributed, again free of charge, to the national governments of industrialised countries.

Polluters subject to a government cap (or funds, banks, or other private or public enterprises) can then also finance carbon saving projects outside the caps and use the resulting extra pollution rights (offsets) in lieu of emissions reduction obligations, or to sell on to third parties, or to speculate with. Thus:

$$CO_2 \text{ reduction under a cap} \quad = \quad \text{offset outside the cap}$$

For example, European Union Allowances (EUAs), the emissions permits traded under the EU cap, are exchangeable with Certified Emissions Reductions (CERs), which are Kyoto Protocol carbon offsets generated in Southern countries outside the European cap:

$$\text{EUA} \quad = \quad \text{CER}$$

Mass production

> Carbon ... will be unambiguously the largest commodity in the world (Richard Sandor 2010).

Such dubious yet versatile equations help market actors mass-produce uniform, tradeable, ownable units of CO_2 reduction (or pollution rights) at the lowest cost for the maximum profit. For example, because carbon dioxide molecules are the same everywhere, it follows that:

$$CO_2 \text{ reduction in place A} \quad = \quad CO_2 \text{ reduction in place B}$$

Carbon businesses under competitive pressure are thus free to choose the cheapest sites for 'reduction production'. Offsets also make possible abstraction from place:

$$CO_2 \text{ reduction in place A} \quad = \quad \text{offset in place C}$$

That is, they take what David Harvey might call the 'spatial fix' of cap and trade (which moves pollution around the 'capped' landscape to wherever it is cheapest to abate) one step further, to territories not covered by caps, notably the global South, where carbon cleanup is even cheaper (Bond 2010). A market in molecular movements also permits indifference to technology type as long as the technologies in question emit the same number of CO_2 molecules:

$$CO_2 \text{ reduction through technology A} \quad = \quad CO_2 \text{ reduction through technology B}$$

Because CO_2 molecules are the same regardless of whether they originate from the burning or decomposition of vegetation or the burning of unearthed fossil fuels, investors can also make use of the equation:

$$\text{a molecule of } CO_2 \text{ of fossil origin} = \text{a molecule of } CO_2 \text{ of biotic origin}$$

in order to benefit from the cost differential between investing in technologies that use less fossil fuels and investing in, say, forest conservation.

Nor need we stop with CO_2. The focus on molecules opens up the cost saving possibility of using other greenhouse gases as well in the climate commodity. Here market construction has benefited from the work of the Intergovernmental Panel on Climate Change (IPCC), which, prompted by the UN's need for national greenhouse gas accounts, has commensurated CO_2 with a range of other greenhouse gases including methane (CH_4), nitrous oxide (N_2O) and various chlorofluorocarbons and fluorocarbons including the industrial by-product HFC-23 (IPCC 1996).

Of course, this requires a lot of fudging. Each gas behaves qualitatively differently in the atmosphere and over different time spans, and the control of each has a different effect on fossil fuel use. Even the IPCC finds itself revising its calculations of the CO_2-calibrated Global Warming Potential (GWP) of various gases every few years, and insists on giving gases different GWPs over 20-year, 100-year and 500-year time horizons. But even such token caveats cannot be accommodated by a market that requires a single, stable number in order to make exchange possible. The UN carbon market, for example, disregards its own IPCC's recent revisions in GWP figures, discards 20-year and 500-year figures, and ignores the often enormous 'error bands' specified by the IPCC (in the case of HFC-23, plus or minus 5 000 CO_2-equivalents).

What remain are the neat equations:

$$CH_4 = 21 \times CO_2$$

$$N_2O = 310 \times CO_2$$

and

$$\text{HFC-23} = 11\,700 \times CO_2$$

In other words, the markets, having abstracted from the climate crisis to CO_2 molecules, now abstract from CO_2 and other gases to a highly simplified carbon dioxide equivalent, or CO_2e, which becomes a new fetish. It becomes even easier to run together, in a seemingly apolitical and self-evident way, activities with different effects on climate history. Thus ex-World Bank executive Robert Goodland (2010), noting that 'domesticated animals cause 32 billion tons of carbon dioxide equivalent, more than the combined impact of industry and energy', effortlessly draws the conclusion that 'replacing livestock products with better alternatives would ... have far more rapid effects on greenhouse gas emissions ... than actions to replace fossil fuels with renewable energy'.

The cost savings achieved by substituting new molecular 'raw materials' for carbon dioxide are considerable, greatly enhancing opportunities for accumulation. For example, burning off just one ton of CH_4 can generate saleable rights to release 21 tons of CO_2 in Europe. Thus more than two dozen giant hog farms operated by Granjas Carroll de Mexico, a subsidiary of the US-based Smithfield Farms, are earning extra revenue by capturing the methane given off by the huge volumes of pig excrement they produce and burning it, and then selling the resulting carbon credits to Cargill International and EcoSecurities (UNEP Risoe Center 2010). Merely by destroying a few thousand tons of HFC-23, similarly, the Mexican chemical manufacturer Quimobasicos is set to sell over 30 million tons of carbon dioxide pollution rights to Goldman Sachs, EcoSecurities, and the Japanese electricity generator J-Power (UNEP Risoe Center 2010).

Assuming that destruction of HFC-23 can be carried out for US$0.25 per ton of CO_2e, and that a ton of CO_2 offset pollution rights can command US$19.50 on the EU ETS spot market (May 2011 prices), both the company and the financial sector intermediaries it sells to can realise super-profits. Industrial buyers of the permits can in turn save US$128.50 per ton by using the rights in lieu of paying fines for not meeting their legal emissions requirements, while industrialists and speculators alike can turn to advantage the US$6 price differential between cheap CERs and more expensive EUAs. Such industrial gas offsets – generated at a handful of industrial installations in China, India, Korea, Mexico, and a few other countries – still account for the bulk of Kyoto Protocol carbon credits, helping to keep carbon pollution rights so cheap that they approach the status of a second 'free allocation' of pollution rights to fossil-intensive European industry. Just as the creation of an abstract, deskilled labour that could be measured in units of time opened up for factory owners from the 18th and 19th centuries onward, the fruitful challenge of how to squeeze the most labour time out of the worker ('moments are the elements of profit', Marx quotes a 19th century British factory inspector saying), so the creation of an abstract climate commodity out of the quantities and movements of CO_2 and other molecules makes possible, for today's climate businesses, a dynamic through which ingenuity and social engineering can be repeatedly exercised to get the most profit out of an assembly line of 'CO_2e reductions'.

Offworld commodity production

A particularly productive element in the cloud of equivalences that make offsets possible is the equation:

$$CO_2\text{e reduction under a cap} = \text{'avoided' } CO_2\text{e emission outside the cap}$$

This equivalence allows offset projects that emit greenhouse gases (and most do) to license the emissions of still more greenhouse gases elsewhere – as long as they emit less than 'would have been released' in the absence of carbon finance. For instance, capped polluters or carbon traders in Europe can purchase carbon pollution rights from coal mining projects in China, provided that the projects burn off some of the methane that seeps out of the mines, on the grounds that by converting methane into carbon dioxide, the projects do less damage to the atmosphere than would have been the case otherwise. Another variant of the equation would allow the forest conservation projects known as REDD (Reducing Emissions from Deforestation and Forest Degradation) to generate carbon credits even if they allowed an *increase* in deforestation, as long as the increase was 'less than would have happened otherwise' (see, e.g., American Carbon Registry 2011).

Thus offset investors make money by, in effect, cleaning up nonexistent extrapolated pollution and taking credit (literally) for its not having become reality. The dirtier that experts can convince regulators that such non-existent extrapolated pollution scenarios are, the more capital can be accumulated, both from pollution rights sales and, on the buyers' side, from delays in investment in no-carbon infrastructure. In fact, through this logic, governments are incentivised to make these imaginary scenarios real by not enforcing or promulgating environmental legislation, since it is by being as dirty as possible that a country creates the most money-making opportunities from carbon markets (see, e.g., Lang 2011). As elsewhere within neoliberal policymaking, the distinction between legal sanctions and market incentives tends to be eroded.

The equation of real and imaginary reductions also requires that 'what would have happened' in the absence of carbon credit sales being determinate and quantifiable in the same way that CO_2e reductions under a cap are determinate and quantifiable.

Counterfactual history, that is, must be given the same epistemic status as actual history and political debate about alternative futures recast as disputes about the correctness of technical predictions.

In a culminating phase of commodity construction, the carbon commodity that has been built up in this way is in turn commensurated with more conventional commodities. It is bundled together with oil and wheat in index funds; deployed as a hedge; used as collateral for international loans (Suppan 2010; Sullivan 2010); and embedded in businesses' long-term investment and profit strategies with regard to energy futures, currency holdings, and so forth.

Exchange, equations, contradictions

> The forms of social practice that gave rise to the new kinds of calculability, and that calculation attempted to format, also continually rendered the world more mobile, uncertain, and incalculable (Timothy Mitchell 2002:118).

'The exchange of commodities', Marx wrote in 1867, 'implies contradictory and mutually exclusive conditions. The further development of the commodity does not abolish these contradictions, but rather provides the form within which they have room to move' (Marx 1990:198). Nowhere is Marx's insight better exemplified today than in the unfolding of the contradictions between use-value and exchange-value traceable in the endlessly proliferating algebra of carbon markets.

The foundational equation 'a better climate = a reduction in CO_2 emissions' starts off the process by eliding the multiple differences between reducing emissions and tackling the climate crisis. The equation obscures, for example, the difference between stepwise molecule reductions over the short or medium term and actions that integrate into a programme that results in fossil fuels being left in the ground permanently. (The industrial slowdown resulting from the financial crisis of 2008, for instance, resulted in more CO_2 emission reductions than all the world's climate markets put together had achieved (Chaffin 2010), yet has not changed structural dependence on fossil fuels.) It also ignores the non-linearity or unverifiability of the relationship – a consequence of the physically 'chaotic', flip-flop nature of the atmospheric system – between any given increment of reduction on the one hand and, on the other, any given increment of climate benefit.

Also left by the wayside is the difference between approaches that focus on molecular movements regardless of their status as 'survival' or 'luxury' emissions (Agrawal and Narain 1991) and approaches that distinguish the two (a choice that has climatic as well as class struggle consequences, since 'survival' emissions tend to have different causes, dynamics, and historical accompaniments than do 'luxury' emissions).

Equating reductions with saleable property rights takes another step away from the climate issue. As fossil fuel use becomes more deeply entrenched through a 'polluter earns' system, the obsession with price discovery draws emphasis away from the long-term structural change demanded by global warming. All things being equal, corporations will choose cheaper alternatives, but if long-term structural alternatives have not been made available, not even the highest prices can compel anyone to choose them; on the contrary, they are likely to incite revolts against the trading system's design. Nor have low prices ever historically been drivers of the kind of structural change that global warming demands. The EU ETS has not incentivised investment away from fossil fuels even in the one sector, electricity generation, that has been consistently short of emissions rights (see, e.g. Deutsche Bank 2009).

Similarly, equating reductions in place A with place B obscures a number of geographically specific factors that make a difference to energy transitions, including the greater influence on technology development that a reduction in emissions from a particular industrial process might have in a high income country, where it is more expensive, than in a low income country (Alfredsson 2009; Jacobson 2009). By abstracting from the tendency for pollution to be concentrated in what in the US are called 'poorer communities of colour,' the equation also bases capital accumulation on historical patterns of class- and racial discrimination, resulting in opposition from networks of underprivileged communities ranging from the California Environmental Justice Movement (California Environmental Justice Network 2010) to India's National Forum of Forest Peoples and Forest Workers (*Mausam* 2008, 2009). Just as the creation and quantification of the working day was a major site and instrument of class struggle in early capitalism, so the assemblage of equations that go into the creation of a climate commodity are major sites and instruments of class struggle in today's nascent carbon markets.

Equating CO_2e reductions that result from different technologies, meanwhile, makes it possible, indeed necessary, to make climatically wrong choices in the name of molecule prices – for example, to use routine, cheap efficiency improvements to delay long-term non-fossil investment, or to build destructive hydroelectric dams that do nothing to displace coal and oil. It also conceals the land-intensive (and thus socially discriminatory) nature of many attempts to 'replace' fossil fuels. Among these are agrofuel schemes in countries such as Brazil and Indonesia, as well as wind power projects such as those in Oaxaca's Tehuantepec isthmus, where many indigenous communities have cheaply signed over land to private wind farm developers from Spain and Mexico – who profit not only from electricity sales but also from using or selling pollution rights in Europe.

The equation:

$$\text{a molecule of } CO_2 \text{ of fossil origin} \;=\; \text{a molecule of } CO_2 \text{ of biotic origin}$$

in addition to weakening the effect of emissions caps by ignoring the difference between the two carbons in terms of climate history, also intensifies climate class struggle in the same way, providing 'scientific' and economic sanction for extensive land grabs from the poor (Dyer and Counsell 2010; Gregersen et al. 2010), who are likely to be displaced at high human cost (not included in the calculations) and see their store of knowledge of low-carbon subsistence livelihoods depleted as a result (also not included in the calculations). The influential 'cost curves' that the equation makes possible also tend to abstract from the difference between forest clearing for commercial agriculture on the one hand and, on the other, rotational forest farming that involves subsequent re-growth of forests and storage of carbon. This abstraction, again, helps pave the way for the deskilling of forest dwellers. As Nathaniel Dyer and Simon Counsell (2010) comment, the 'argument that we need a new economic model to account for [climate change] externalities and to put our economies on a sustainable path' has ironically led to cost curves which, with their 'hidden costs and partial analysis', are 'similar to the narrow economic approach that contributed to the problem that we are now attempting to solve'.

Thus Aritana Yawalapiti, an indigenous leader in the upper Xingu region of Brazil, reported in November 2010 that REDD carbon forestry promoters visiting his territory had told his community that they would have to reduce forest burning if they were to be paid for producing carbon pollution licenses. But, Aritana objected,

> We always burn at a place where we fish, hunt or open a small farmland area … we open a space to farm, we plant, we collect manioc, after some years everything recuperates again … the forest grows back, while we plant at another place (Sommer 2010).

As banners carried by Karen indigenous people from Thailand during a demonstration outside the Bangkok climate negotiations in 2009 explained, 'people who live with the forest don't want REDD … we conserve forests because forests are life, not a commodity'. Here as elsewhere, accumulation in the carbon markets takes place not through 'decarbonisation' but through dispossession.

The crucial 'offset' equation:

$$\text{actual } CO_2e \text{ reduction} \quad = \quad \text{counterfactual } CO_2e \text{ reduction}$$

conceals and engenders yet other forms of the contradiction between use-value and exchange-value. By making accumulation dependent not only on finding or postulating, but also (if possible) on creating as much greenhouse gas as possible so that it can then be 'avoided', applications of the equation continually generate perverse consequences. This phenomenon is most visible with respect to HFC-23 and N_2O, manufacturing by-products whose cleanup now often generates more profit for their manufacturers than the primary products of the processes in question (Pearce 2010); but the phenomenon is general. As corporations are given incentives not to obey environmental laws and governments are given incentives not to promulgate or enforce them, opposition grows from environmental protection movements in countries such as the Philippines, South Africa, and Nigeria, where oil companies gain extra profit for supposedly avoiding gas flaring activities that are illegal and unconstitutional anyway (Osuoka 2009; Docena 2010).

Also provoking resistance is the carbon market's general methodological imperative of isolating a single counterfactual story line as a baseline for offset savings: in treating history deterministically with the exception of the activities of carbon financiers and offset project managers, this technical requirement for a market 'eliminates' the ability of workers and farmers to create their own history (FASE 2003; Gilbertson and Reyes 2009). Here, too, the internalisation of the global warming externality in carbon prices gives rise to fresh externalities, as the contradiction between use-value and exchange-value is re-activated at each step and calculative technologies continually undermine their own efficacy (compare Soros 2008). As with the market in complex financial derivatives, the more extensive and liquid the commodity trade in question becomes, the less successful it is in meeting its ostensible 'objective' (Lohmann 2009b). Climate change, no less than price uncertainties, has proved to be a singularly recalcitrant subject for the headlong, overconfident efforts at commodification characteristic of the neoliberal era.

Regulation and decommodification

Over the last decade-and-a-half, the overwhelming contradictions of climate markets, while shape-shifting continually, have only grown more intense as new equations proliferate and market actors, regulators, biogeochemical systems, forests, technological complexes, and grassroots resistance networks each make their plays. The strategic question in the face of this dynamic is how to build the most effective possible movements to address the climate threat that is now posed by carbon markets themselves.

This cannot be done through any type of regulation (or, to use the preferred neoliberal term, 'governance') that does not contribute to movements for decommodification of the earth's carbon-cycling capacity. For example, no additional equivalences, surveillance procedures, or technical criteria for determining when a carbon offset project goes beyond 'business as usual' could ever relieve the contradictions built into the equation:

$$\text{actual } CO_2\text{e reduction} = \text{counterfactual } CO_2\text{e reduction.}$$

Current attempts to do so are merely giving underlying contradictions described in this article more room to move. The effect has been to reinforce the supply-side dominance in the offset markets of large polluting corporations that operate in the global South – Sasol, Mondi, Rhodia, Tata, Birla, Jindal, and the like (UNEP Risoe Center 2010) – who are better able than others to devote resources to navigating the growing regulatory and planning mazes in the service of gleaning new revenues for activities that reinforce fossil fuel use.

To take another example, the regulatory principle according to which development projects must obtain the free prior informed consent of affected communities becomes virtually useless once 'climate mitigation' has been globalised through carbon offset projects. A forestry project in Australia selling pollution licenses to ConocoPhillips, for example, would have to obtain the consent not only of the affected community in Australia, but also of communities affected by Conoco operations in the US, Libya, Peru, Vietnam, Kazakhstan, Greenland, and other parts of the world – clearly an impractical requirement.

In this context, to interpret the analysis of the equations sketched in this paper as a demand for them to be 'fixed' and 'elaborated' would be delusional. Carbon market regulation that relies on 'more and better' internalisation merely creates dangerous new externalities, making climate policy even more counterproductive than it already is.

Only those regulations that limit or reduce commodification have much of a chance of limiting the damage carbon markets do, or of playing a (small) part in the longer-term project of forcing policy makers to phase out carbon trading. As is suggested by the taxonomy of equations discussed in this article, commodification and decommodification have many forms and degrees, and even governments that work within a carbon-trading framework are sometimes induced to undertake actions with modest decommodification effects. For example, the EU has recently decided to stop applying the equation:

$$\text{HFC-23} = 11\,700 \times CO_2$$

by banning HFC-23 credits from sale as of 2013. The reasons for this move are complex, involving not only recent scandals over the issuance of a flood of blatantly bogus pollution rights from industrial gas projects (EIA 2010), but also fears that European industries in the sector in question may relocate to the global South to take advantage of offset revenues; a desire to reduce transaction costs in the manufacture of carbon offsets by sourcing them from entire sectors rather than individual projects; and worries that an oversupply of carbon credits will undermine market operations. Nevertheless, the curb does demonstrate the possibility of rolling back commodification rather than extending it, as do environmentalist campaigns to abolish offsets and 'deactivate' equations such as:

$$\text{EUA} = \text{CER} = \text{AAU}.$$

More generally, regulators sometimes come under pressures that result in their taking *ad hoc* actions that have a temporary decommodification effect, for example, by restricting:

- Who can sell or buy carbon commodities and for what reason – for instance, moves to restrict access of speculators to the markets;
- what traders can sell or buy – for instance, moves to get rid of N2O or coal-project credits, restrict the exchangeability of allowances, or limit securitisation;
- where they can sell or buy the commodity – for instance, restrictions on over-the-counter trading or on trading technology;
- when they can sell or buy – for instance, limits on banking and borrowing; and
- how traders can sell or buy – for instance, restrictions on the velocity of trading.

However, moves that reduce carbon markets' liquidity and factitious 'efficiency', as well as the opportunities they provide for the financial sector, are bound to continue to rouse the opposition of many powerful actors in the carbon markets as well as some market architects. Discussions are underway at the UN, moreover, to scale up commodification by allowing whole sectors to produce carbon credits by performing 'better than they would otherwise.'

In the end, there can be no substitute for strategic popular movements around decommodification of climate benefit in alliance with related movements against commodification of water, electricity, health services, and fossil fuels and in support of land rights, labour, tax reform, alternative energy, alternative transport, food sovereignty, and public control of the financial sector. Such alliances are already contributing to a shift in the centre of gravity of climate change activism away from the technocratic programmes advocated by governments, corporations, and large environmental NGOs, which have always been organised around molecule flows, temperature targets, and market mechanisms. Some of the most powerful voices for radical climate action and against carbon markets to emerge recently include the international peasant movement La Via Campesina; movements against fossil fuel extraction in countries such as Nigeria, Canada, and Ecuador; various Indigenous Peoples' networks; environmental justice movements among the poor in cities such as Los Angeles and Durban; labour unions in countries such as Mexico; and so forth.

As such connections suggest, the class and race dimensions of the struggle over the equations analysed in this article are bound to remain central. During the last days of the December 2010 climate summit in Cancún, UN security ordered non-governmental delegates who were displaying anti-REDD stickers to remove them and temporarily suspended the accreditation of Tom Goldtooth of the Indigenous Environmental Network (IEN), a prominent opponent of REDD (*Democracy Now!* 2010), as well as that of representatives of low-income communities of colour in California who were also critical of carbon trading. 'I came to [the summit] representing the public health concerns of low-income communities of colour living in Los Angeles being impacted by toxic emissions,' said Sunyoung Yang of the Los Angeles Bus Riders Union. 'Throughout the past two weeks I have seen how the … meeting has systematically limited and suppressed voices of dissent to programmes being promoted through the UN such as REDD which will only increase the poisoning of the communities I represent back home' (Global Grassroots Justice Alliance et al. 2010).

Left unharrassed, by contrast, were demonstrators and delegates from the network 350.org, who had staged larger and noisier protests demanding a target of 1.5 degrees Celsius of warming as well as a limitation of atmospheric concentrations of CO_2 to 350 parts per million, but have refrained from critiquing market approaches and have embraced the CO_2 fetish (indeed, the name of their organisation embodies it). As the Indian activist Soumya Dutta noted following the conference, 'Only questions and protests which were clear in their political message against present-day big-money interest were pro-actively silenced' (Dutta 2010).

Conclusion: Internalisation as externalisation

Carbon markets are a particularly disastrous example of what can happen when the cluster of processes commonly associated with neoliberalism is let loose on environmental crises. But the lessons are more general. Commodity solutions always reinterpret and transform the social and environmental challenges that they confront. Their goals are never exogenous but are incessantly reshaped by the very process of addressing them. Hence the 'internalisation of environmental externalities' associated with market environmentalism is better conceived not as a (successful or failed) attempt at 'environmental problem-solving', but rather as a continuous changing of the subject.

In order to be 'internalised', environmental harms of any complexity must be simplified, reformatted, made abstract, quantifiable, and transferrable in a process that obscures many of their characteristics while introducing fresh problems. For instance, turning biodiversity into a commodity means transforming it into measurable units, inevitably divorcing it from the human and nonhuman context in which it has evolved while abstracting from habitat, species, genus, or the like. By the same token, conjuring up bankable 'wetlands credits' capable of expanding opportunities for the circulation of capital involves 'ignoring a great deal of ecological information' (Robertson 2004), setting off conflicts among bankers, regulators, and scientists alike.

In the process of developing skills at internalisation, moreover, internalisers set aside, lose, or destroy other skills, theirs and others' (Lohmann 2009a; compare Mirowski 2011). In order to be priced low enough to be traded, 'environmental service' commodities also unavoidably conceal the human labour, study, and other activities that go into creating, maintaining, and restoring ecosystems. Competition among 'environmental service' businesses only increases pressures to disaggregate and decontextualise in the service of cost reductions. The harms associated with the simplified and restructured socio-natures that result – for instance, landscapes in which regions of accelerated degradation are interspersed with areas specialising in the production of a few technician-selected aspects of 'nature' – must then also be 'internalised', whether through the creation of still further commodities or through another layer of expert 'governance.' And so on.

Any process of internalisation, in short, creates its own externalities. Internalisation itself is externalisation (Callon 1998). The question is not whether but when a new internalisation will be required or enabled; each act of internalisation gives rise to a need or a possibility for yet further internalisations. Neither a coherent strategy for 'making capitalism sustainable' nor mere opportunistic greenwash for supposedly more fundamental processes, internalisation is merely one more example of a long-familiar process through which business treats the crises it helps to throw up as a basis for further accumulation.

References

Agrawal, A. and S. Narain. 1991. *Global warming in an unequal world: A case of environmental colonialism*. New Delhi: Centre for Science and Environment.

Alfredsson, E. 2009. Perspectives on cost efficiency and technological change. Presentation at Swedish Society for Nature Conservation. Stockholm. September 12.

American Carbon Registry. 2011. Methodology for REDD: Avoiding Planned Deforestation. Arlington: Winrock International, http://www.americancarbonregistry.org/carbon-accounting/ACR%20Methodology%20for%20REDD%20-%20Avoiding%20Planned%20Deforestation%20v1.0%20April%202011.pdf.

Arthur, W.B. 1999. *Increasing returns and path dependence in the economy*. Cambridge: Cambridge University Press.

Ball, J. 2008. Pollution credits let dumps double dip: Landfills find new revenue in trading system meant to curb greenhouse emissions. *Wall Street Journal*. October 20.

Bond, P. 2010. Climate justice politics across space and scale. *Human Geography* 3 (2): 49–62.

Brinkley, A. and S. Less. 2010. *Carbon omissions: Consumption-based accounting for international carbon emissions*. London: Policy Exchange.

Broughton, E. and R. Pinard. 2011. *What's in a name? Market-based instruments for biodiversity*. Paris: Institut du Developpement Durable et des Relations Internationales.

Buiter, W. 2007. Carbon offsets: Open house for waste, fraud and corruption, http://blogs.ft.com/maverecon/2007/07/carbon-offsets-html.

California Environmental Justice Movement. 2010. *EJ Matters* website, http://www.ejmatters.org/.

Callon, M. 1998. An essay on framing and overflowing: Economic externalities revisited by sociology. In M. Callon, ed. *The laws of the markets*, 244–269. Oxford: Blackwell.

CDM & JI Monitor. 2011. EB dodges coal decision, opens CDM up for poorer nations. July 27.

Carr, M. and S. Lomax. 2010. Richard Sandor exits carbon trading with sale of climate exchange to ICE. *Bloomberg*. April 30.

Chaffin, J. 2010. Economic crisis cuts European carbon emissions. *Financial Times*. April 1.

Chichilnisky, G. 2009. *Saving Kyoto: An insider's guide to how it works, why it matters and what it means for the future.* London: New Holland.

Climate Action Network (CAN). 2010. CAN-International submission on the development of standardized baselines. March, http://cdm.unfccc.int/about/standardized_baselines/sbase_cani.pdf.

Cole, D. H. 2002. *Pollution and property: Comparing ownership institutions for environmental protection.* Cambridge: Cambridge University Press.

C-Quest Capital. 2011. Ken Newcombe, CEO. http://www.cquestcapital.com/team.shtml.

Democracy Now!. 2010. Prominent Indigenous environmental activist blocked from UN climate talks. December 9, http://www.democracynow.org/2010/12/9/prominent_indigenous_environmental_activist_blocked_from.

Dennett, D.C. 1995. *Darwin's dangerous idea: Evolution and the meanings of life*. New York: Simon & Schuster.

Deutsche Bank. 2009. The long and short of it: Power sector key to EUA & CER prices. Carbon Emissions Commodities Report. London: Deutsche Bank. May 5.

Docena, H. 2010. *The CDM in the Philippines: Costly, dirty, money-making schemes.* Bangkok: Focus on the Global South.

Driesen, D. 2003. *The economic dynamics of environmental law*. Cambridge, MA: MIT Press.

Dutta, S. 2010. Email communication to Climate Justice Now!. December 12.

Dyer, N. and S. Counsell. 2010. *McREDD: How McKinsey 'cost-curves' are distorting REDD.* London: Rainforest Foundation.

Environmental Investigation Agency (EIA). 2010. Q&A on industrial gases and the CDM in the E.U. ETS. London: EIA, http://www.eia-international.org/files/reports212-1.pdf.

FASE. 2003. Open letter to executives and investors in the Prototype Carbon Fund. Espiritu Santo: FASE. May 23.

Figueres, C. 2011. Curriculum vitae, *http://www.figueresonline.com/CFO_English_Long.pdf.*

Foster, J.B. 2003. A planetary defeat: The failure of global environmental reform. *Monthly Review* 54 (8).

Gilbertson, T. and O. Reyes. 2009. *Carbon trading: How it works and why it fails.* Uppsala: Dag Hammarskjold Foundation.

Global Grassroots Justice Alliance and Youth for Climate Justice. 2010. Statement on being expelled from COP 16. Montevideo: World Rainforest Movement, http://www.wrm.org.uy/COP16/Statement_from_the_expelled.html.

Goodland, R. and J. Anhang. 2010. Livestock and climate change. *Worldwatch*. November/December: 10–19.

Gregersen, H., H. El Lakany, A. Karsenty, and A. White. 2010. *Does the opportunity cost approach indicate the real cost of REDD+? Rights and realities of paying for REDD+*. Washington: Rights and Resources Initiative.

Gronewold, N. 2011. UN body relaxes rules to spread clean development credits around. *Environment & Energy Daily*. July 25.

Harvey, D. 2010. *The enigma of capital and the crises of capitalism*. London: Profile Books.

Helm, D. 2010. Government failure, rent-seeking, and capture: The design of climate change policy. *Oxford Review of Economic Policy* 26 (2): 182–196.

Hulme, M. 2009. *Why we disagree about climate change: Understanding controversy, inaction and opportunity*. Cambridge: Cambridge Univ. Press.

IDEAcarbon. 2011. Empowering global carbon markets. London: IDEAcarbon, http://www.ideacarbon.com/.

Intergovernmental Panel on Climate Change (IPCC). 1996. *Climate change 1995: The science of climate change*. Cambridge: Cambridge Univ. Press.

International Emissions Trading Association (IETA). 2010. IETA response to the call for input on modalities and procedures for standardized baselines. March 22, *http://cdm.unfccc.int/about/standardized_baselines/sbase_ieta.pdf*.

Jacobson, M.Z. 2009. Effects of local CO_2 domes and of global CO_2 changes on California's air pollution and health. Testimony before the Environmental Protection Agency Hearing on the California Waiver, Washington. March 5.

Kysar, D. 2010. Not carbon offsets, but carbon upsets. *The Guardian*, August 29.

Lang, C. 2011. McKinsey's advice on REDD is 'fundamentally flawed,' says Greenpeace. *REDD-Monitor*. April 8, http://www.redd-monitor.org/2011/04/08/mckinsey-advice-on-redd-is-fundamentally-flawed-says-greenpeace/.

Lohmann, L. 2009a. Toward a different debate in environmental accounting: The cases of carbon and cost-benefit. *Accounting, Organizations and Society* 34 (3–4): 499–534.
———. 2009b. *When markets are poison: Learning about climate policy from the financial crisis*. Sturminster Newton: The Corner House.
———. 2011. *Mercados de carbono: Neoliberalizacion del clima*, Quito: Abya Yala.

McGarrity, J. 2011. CO_2 market exodus to grow amid ETS gloom, policy uncertainty. *Point Carbon*, July 20.

Marx, K. 1990. *Capital Vol. 1*. Trans. Ben Fowkes. London: Penguin.

Mausam: Indian Climate Change Magazine. 2008. Kolkata: NESPON, http://www.thecornerhouse.org.uk/sites/thecornerhouse.org.uk/files/Mausam_July-Sept2008.pdf.

———. 2009. Kolkata: NESPON, http://www.thecornerhouse.org.uk/sites/thecornerhouse.org.uk/files/Mausam2-5.pdf.

Mirowski, P. 2011. *Science-mart: Privatizing American science*. Cambridge, MA: Harvard University Press.

Mitchell, T. 2002. *Rule of experts: Egypt, techno-politics, modernity*. Berkeley: University of California Press.

Newell, P. and M. Paterson. 2010. *Climate capitalism: Global warming and the transformation of the global economy*. Cambridge: Cambridge University Press.

Nicholls, M. 2011. Energy efficiency directive raises fears for E.U. ETS. *Environmental Finance*, July-August.

Osuoka, I. 2009. Paying the polluter? The relegation of local community concerns in 'carbon credit' proposals of oil corporations in Nigeria. Unpublished manuscript. April.

Parekh, P. 2011. Close the crediting loophole in coal-fired projects in CDM. *CDM & JI Monitor.* July 27.

Pearce, F. 2010. Carbon trading tempts firms to make greenhouse gas. *New Scientist*. December 16.

Robertson, M.M. 2004. The neoliberalization of ecosystem services: Wetland mitigation banking and problems in environmental governance. *Geoforum* 35 (3): 361–373.

Sandbag. 2011. *Buckle up: Tighten the cap and avoid the carbon crash.* London: Sandbag.

Sandor, R. 2010. Interview with Rick Santelli. Chicago: *CNBC*. April 26.

Sills, B. 2011. Global carbon credits die as smart money embraces India's RECs. *Bloomberg Markets*. March 22.

Sommer, R. 2010. Pirakuma Yawalapiti, Xingu spokesperson, about carbon trading. Video interview. Xingu River, Brazil, http://www.youtube.com/watch?v=_JSM6gaM9CA, http://www.youtube.com/watch?v=JMs3szvzfeA&feature=related.

Soros, G. 2008. *The new paradigm for financial markets: The credit crisis of 2008 and what it means*. New York: Public Affairs.

Stern, N. 2009. *The global deal: Climate change and the creation of a new era of progress and prosperity.* New York: Public Affairs.

Sullivan, S. 2010. The environmentality of 'Earth Incorporated': On contemporary primitive accumulation and the financialization of environmental conservation. Paper presented at the conference An Environmental History of Neoliberalism, Lund University. May 6–8.

Suppan, S. 2010. *Trusting in dark (carbon) markets? The UN high-level advisory group on climate finance*. Minneapolis: Institute for Agriculture and Trade Policy.

The Munden Group. 2011. *REDD and forest carbon: Market-based critique and recommendations.* New York: Munden Group.

United Nations Environment Program (UNEP) Risoe Centre. 2010. *CDM Pipeline* (spreadsheet), http://cdmpipeline.org/.

United Nations Framework Convention on Climate Change (UNFCCC). 2011. Call for public inputs on sustainability benefits, http://cdm.unfccc.int/public_inputs/2011/sustainability_benefits/index.html.

Unruh, G.C. 2000. Understanding carbon lock-in. *Energy Policy* 28: 817–830.

Weitzman, M. 2008. On modeling and interpreting the economics of catastrophic climate change. REStat FINAL Version, July 7, 2008, http://www.economics.harvard.edu/faculty/weitzman/files/REStatFINAL.pdf.

World Bank. 2011. *State and trends of the carbon market 2011.* Washington: World Bank.

Wynn, G. and N. Chestney. 2011. Carbon offsets are worst performing commodity: Face rising supply at time of economy fears. *Reuters*. August 5.

Endnote

[1] Thanks for useful discussion to Oscar Reyes, Steve Suppan, Andres Barreda, Jutta Kill, Ricardo Coelho, Hendro Sangkoyo, Martin Bitter, Esperanza Martinez, Ivonne Yanez, Matthew Paterson, Silvia Ribeiro, Raul Garcia, John Saxe Fernandez, Herbert Docena, Patrick Bond, John O'Neill, Erik Swyngedouw and Mark Schapiro. An earlier version of this article appeared as *Un álgebra interminable: las contradicciones de los mercados climáticos* in Lohmann 2011.

Chapter 9

Zombie Carbon and Sectoral Market Mechanisms

Oscar Reyes[1]

The global carbon market is in crisis. Proposed emissions trading schemes in the USA, Japan, and Canada have stalled indefinitely; new markets in Australia and South Korea face significant delays; and climate justice activists have successfully blocked the start of a planned scheme in California. Trading has become ever more concentrated around the EU Emissions Trading System (EU ETS), which could well see carbon permit prices drop to zero if the 27-country bloc adopts stricter guidelines on energy efficiency (Harrison 2011). Overall carbon trading volumes were lower in 2010 than in the previous year. The Clean Development Mechanism (CDM), the carbon offsetting scheme at the heart of the Kyoto Protocol, has declined for four years running, with fewer credits purchased from new projects than at any time since the Protocol came into force in 2005 (World Bank 2011a). The price of CDM credits continues to fall, and they are now 'the world's worst performing commodity' (Wynn and Chestney 2011).

Perhaps confusing these contractions for birth pangs, the EU and the World Bank are pushing to create new international carbon market mechanisms in the context of United Nations Framework Convention on Climate Change (UNFCCC) international climate negotiations. In particular, there is a focus on creating new 'sectoral' carbon markets, which would move beyond the project-by-project basis of the existing CDM and issue carbon allowances in relation to whole economic sectors.

The UN Climate Change Conference in Cancún (COP 16) agreed that one or more of these new market-based mechanisms should be established at COP 17 in Durban. The industrialised countries advancing these proposals want to see them expand upon (and partially replace, in the case of middle-income countries) the CDM, providing a legal framework for continuing the carbon market experiment that began with the Kyoto Protocol, even in the absence of any new 'emissions reductions' pledges lodged under that treaty.

With global climate talks showing few signs of progress, the debate on establishing new carbon markets may roll on well past the Durban conference. But the World Bank, in particular, is continuing to push for the expansion of these markets regardless of the Durban outcome, while a series of bilateral initiatives are emerging alongside the multilateral framework. This article focuses on 'sectoral' carbon markets, explaining and critiquing two proposals that are central to the expansion of new carbon market mechanisms. It highlights the role of the World Bank's Partnership for Market Readiness (PMR) as a catalyst for new carbon markets in middle-income countries and uses the same mechanism as a window through which to explore the rise of bilateral initiatives. It concludes that neither failures at the level of international negotiations nor a slump in carbon credit prices have yet stopped the attempts of the EU and World Bank, in particular, to roll out new carbon market mechanisms.

Going sectoral

Deadlock in Durban seems the most likely outcome of climate talks, with the debate on the 'legal form' of any agreement taking centre stage. There are currently two tracks within the UNFCCC negotiations. An Ad Hoc Working Group on Further Commitments for Annex 1 Countries under the Kyoto Protocol (AWG-KP) was established in December 2005. This was later joined by the Ad Hoc Working Group on Long Term Cooperative Action (AWG-LCA, hereafter LCA), which takes the Bali Action Plan (BAP) of December 2007 as its starting point.

The 'legal form' debate centres on whether or not to unify the results of these tracks into a single, post-Kyoto treaty. This is an arcane framing for a dispute that is ultimately about power and equity: who should take on responsibility for reducing greenhouse gas emissions, and can states be held to account if they backtrack on their commitments? These are far from theoretical considerations: the USA wrote carbon markets into the 1997 Kyoto Protocol but then famously failed to ratify that treaty. It has now been joined by Canada (which is almost certain to miss its Kyoto target), Japan, and Russia in rejecting any continuation of the Protocol after 2012. These industrialised countries want to 'kill Kyoto', their aim being to get rid of internationally-binding climate targets, while keeping hold of the carbon markets. Establishing new carbon market mechanisms outside of the Kyoto Protocol is part of this package.

To this end, the debate on these new mechanisms mainly rests within the LCA and is being elaborated under Section 1(b)(5) of the 2007 Bali Action Plan, which calls for '[v]arious approaches, including opportunities for using markets, to enhance the cost-effectiveness of, and to promote, mitigation actions' (UNFCCC 2007a). The Cancún Accords resulting from COP 16 ask that new market and non-market mechanisms be considered for agreement in Durban. A short consultation was held in early 2011, and consolidated by the LCA Secretariat into a 'Synthesis report on information on various approaches in enhancing the cost-effectiveness of, and promoting, mitigation actions' (hereafter, *Synthesis Report*), which forms the basis of ongoing negotiations.

Various proposals remain on the table, but they centre on sectoral carbon markets, following considerable efforts by the European Commission, in particular, to promote them. These come in two main variants: 'sectoral crediting' and 'sectoral trading'. The former would issue credits for reductions in pollution relative to a projected baseline, based on measurements over an agreed time period. The latter would issue 'permits to pollute' up front in relation to a target level of emissions.[2]

The new schemes are presented as distributing the 'benefits' of carbon finance more evenly in the South, but it is likely that the main winners would be large (often transnational) corporations operating in middle-income countries. The reasons for this relate to the market fundamentals of key economic sectors as well as inequalities of information and capacity to engage in non-project-based trading systems. Beyond the development financing rhetoric, an intended effect of sectoral markets is to draw so-called major emitters into climate change mitigation obligations, and so redress perceived trade imbalances resulting from measures to tackle greenhouse gas emissions. Climate negotiations have long been treated as trade negotiations, so this should come as little surprise. More noteworthy, perhaps, is the increasing tendency to double count carbon market revenues as sources of 'climate finance'. It forms part of a growing tendency to conceive of such financial flows as little more than leveraging international private finance – a polite way to describe corporate welfare schemes with little visible benefit (and the potential for considerable harm) to the majority of people in Southern countries.

Sectoral crediting

The *Synthesis Report* explains sectoral crediting in the following terms:

> On a crediting basis, a reference level for emissions within a boundary (e.g., a sector) would be determined, possibly at a level below "business as usual." The emissions within this boundary would then be monitored during a period of time, known as a "reference period." If, at the end of the reference period, actual emission levels were below the reference level, a quantity of credits corresponding to the difference would be issued. These credits would then be distributed, through a method to be determined, among the emitters within this boundary. The emitters would therefore have an incentive to limit their emissions, as this would enable them to gain credits that could be monetized. A variant of this proposal would be that credits are issued for emissions that are avoided during the period (UNFCCC 2011a).

This requires a little decoding, starting with the obvious question of what activities would be covered. Although the precise list of sectors is not yet fixed, it is generally understood to include manufacturing sectors exposed to international competition – including steel, cement and lime, pulp and paper, aluminium, and 'upstream' oil- and gas-production emissions (e.g., from gas venting and flaring) – as well as the power sector. Such crediting may also include economic sub-sectors, such as public transport (UNFCCC 2011b:52,45). Separate discussions are being held within the context of UNFCCC negotiations regarding the possibility of new carbon market mechanisms involving deforestation and agriculture.

Emissions trading is also under consideration in the aviation and shipping sectors, although these discussions are currently being conducted under the mandate of the International Civil Aviation Organization (ICAO) and the International Maritime Organization (IMO), institutions that have avoided binding agreements to control climate change for more than thirteen years.[3]

The *Synthesis Report* is deliberately vague on how emissions would be counted and credits issued, although the baseline-and-credit system on offer is similar in many fundamental aspects to the CDM. A baseline or reference level is a projection of future emissions, which incorporates assumptions about past polluting practice, future economic growth, and the likely trajectory of technology development. Minor alterations to these assumptions can result in major differences in how many credits are issued and, therefore, how much money is generated by the scheme.

One of the most controversial aspects of the proposed mechanism is the suggestion that crediting should start at a threshold that is 'possibly at a level below "business as usual."' This would require developing countries to significantly alter their emissions trajectory *before* carbon credits are issued. In this sense, the scheme is said to advance beyond the 'zero sum' game of the CDM, where all of the credits issued simply prevent more expensive actions to reduce greenhouse gas emissions in industrialised countries (Lazarowicz 2009:61).

The move 'beyond offsetting' is far from eliminating offsetting or any of the problems associated with it, however. In covering whole economic sectors instead of individual projects, sectoral crediting would increase the volume of carbon offsets generated; in fact, that is one of its main goals.

No penalty is incurred by the developing country if this threshold is not met, which is why the scheme is sometimes referred to as involving a 'no-lose' target. But no credits are issued in that case, either. Credits are issued to the developing country if the emissions threshold is exceeded. These can then be sold back to industrialised countries (and companies covered by emissions trading schemes within these countries) as offsets. In other words, these credits are bought so that companies and governments in industrialised countries do not have to reduce their emissions at source.

Limiting the flow of offsets by means of a 'crediting threshold' is, in part, a measure to 'ensure that the supply of credits does not overwhelm demand' (Baron et al 2009, 16). It is, moreover, a proposal that has the potential to displace more of the costs of addressing climate change onto the developing countries, collectively. If the autonomous efforts of developing countries are linked to an offset market, it is feared that low-cost emissions reduction potential would be counted towards Annex I country targets rather than as part of the effort by developing countries to advance cleaner development paths (Third World Network 2009).

Sectoral trading

Sectoral trading is a type of cap and trade scheme. It aims to apply a cap (or limit) on greenhouse gas emissions relating to a particular economic activity. Companies are issued licenses to pollute (carbon permits or emissions allowances) and can then choose to cut their emissions or buy permits from others that have a surplus ('a trade').

There are two key differences between sectoral trading and sectoral crediting. First, the issuance of allowances happens at the start of the period of trading rather than at the end. If an agreed overall target is not met, the government or companies covered by the scheme would have to purchase extra carbon allowances from abroad. With sectoral crediting, by contrast, credits would only be issued at the end of a period of trading if emissions were below an agreed baseline (Lazarowicz 2009:57).

Second, sectoral trading cannot be operated on a voluntary basis. Or, rather, the voluntary aspect relates solely to whether countries choose to opt into and create such a scheme in the first place. Once created, the target must be a 'defined absolute target' if some scarcity is to be maintained, which is a requirement for the permits to retain their monetary value. In this sense, the scheme forms part of an attempt to weaken the differentiation within the global climate regime between Annex 1 (developed) countries, which take on binding emissions targets, and non-Annex 1 (developing) countries, which do not.

The question of how sectoral trading allowances would be allocated, or by whom, is left unanswered – although the experience of existing cap and trade schemes shows a consistent pattern of over-allocation (the number of allowances exceeds the cap, so no pollution is limited) and has tended to involve free handouts of permits to pollute. Nor is it yet established whether the sectors covered would be defined on a sub-national, national, regional, or global basis – although the *Synthesis Report* claims that parties have expressed a preference for national or bilateral systems (UNFCCC 2011a:7).

The World Bank's push for new carbon markets

The UNFCCC climate negotiations are not the only means of advancing new carbon market mechanisms, however. A growing number of initiatives are pushing beyond the official UN framework, in response to the sustained deadlock within the talks. The World Bank is at the vanguard of these measures, and although its Partnership for Market Readiness (PMR) involves a relatively modest financial outlay, its potential to catalyse new markets (and serve as a bellwether for national and bilateral initiatives) makes it worthy of critical attention.

Launched at the UN Climate Change Conference in Cancún (COP 16), the PMR is a new carbon fund 'aimed at major emerging economies and middle-income countries interested in exploring new carbon market mechanisms, including sectoral crediting mechanism[s]' (European Commission 2010a:19).

A range of economic activities might eventually be covered by these new markets, with the current (non-exhaustive) list including power generation, iron and steel, transport, construction/buildings, cement, energy efficiency, waste management, and initiatives towards 'low-carbon cities' (World Bank Climate Finance Unit 2011a).

Promoting 'market readiness' is strategically important for the Bank (and its financial backers) in attempting to open up new forms of carbon markets in countries which until now have not been obliged to monitor their emissions. This activity pre-empts a political decision within international climate negotiations, despite attempts to dress up the intervention as a merely technical exercise. As the European Commission notes,

> Some countries may perceive the [PMR] project as potentially jeopardizing their negotiation positions and the process under the UNFCCC. However, such risk could be mitigated by focusing on the technical discussions and on-the-ground capacity building (European Commission 2010a:23).

This is by no means the first time the Bank has made such a move. At the UN Climate Change Conference in Bali in December 2007 (COP 13), the World Bank launched its Forest Carbon Partnership Facility (FCPF), a 'market readiness' initiative for Reducing Emissions from Deforestation and Forest Degradation (REDD). As Benoit Bosquet, the Bank official who led the development of the facility, put it at the time, 'The facility's ultimate goal is to jump-start a forest carbon market.' This despite the lack of any UN agreement on REDD carbon markets.

Moreover, the Bank clearly intends to pursue the creation of new carbon market mechanisms irrespective of UNFCCC negotiations. The European Commission points out that '*Regardless if the final decision on the establishment of new carbon market mechanisms will be taken under the auspices of the UNFCCC or via bilateral or multilateral agreements*, demonstration actions like the PMR will improve understanding on the options for practical implementation of new and scaled-up carbon market mechanisms' [emphasis added] (European Commission 2010a:23).

As of June 2011, the Partnership had approved initial grants of US$350 000 to Chile, China, Colombia, Costa Rica, Indonesia, Mexico, Thailand, and Turkey (World Bank Climate Finance Unit 2011b). Each of the eight countries will now develop a 'Market Readiness Proposal' to detail their plans. Two further countries, Morocco and Ukraine, have been confirmed as participants, while Brazil, Jordan, South Africa, and Vietnam are at various stages of submitting 'Expressions of Interest' (World Bank Climate

Finance Unit 2011c:34). The initial projects include setting up a carbon offset registry in Mexico and establishing regional pilot schemes for emissions trading in the power sector in China. The largest share of the money will be allocated to creating systems for Monitoring, Reporting and Verifying (MRV) (European Commission 2010a:25).

Funding pledges so far have come from Australia, the European Commission, Germany, Japan, Norway, The Netherlands, Spain, Switzerland, the United Kingdom, and the United States (World Bank Climate Finance Unit 2011b). A little under US$70 million has been pledged to date, with all or most of the money coming from 'fast start financing,' the package of 'urgent' measures announced as part of the 2009 Copenhagen Accord (Australian Government 2010:10; World Bank Climate Finance Unit 2011b; European Commission 2010a:18).

A closer analysis shows that the most significant source of money for the new mechanisms proposed by the Fund will come from the countries that are supposed to be its beneficiaries. As the European Commission explains, each beneficiary country will initially be allocated US$200 000 to identify relevant sectors for the scheme. An average of US$5 million will subsequently be spent on programme implementation in each participating country, US$3 million of which will be dedicated to establishing systems for data collection, monitoring, and reporting (European Commission 2010:23). In this regard, the programme closely follows the format adopted in the development of 'REDD-readiness' initiatives, such as the World Bank Forest Carbon Partnership Facility.

However, as the European Commission points out, 'US$5 million is not sufficient to bring [the] PMR programme to do piloting. The beneficiary countries will be required to allocate human and financial resources to perform all above-mentioned tasks' (European Commission 2010a:26). The scale of this shortfall can be seen when the PMR figures are compared with estimates that appear in a 2009 study commissioned by the UK Office of Climate Change Global Carbon Trading Project. The comparison is especially noteworthy, because it was conducted by Ecofys, a consultancy that is one of the main advisors to the World Bank and OECD on 'market readiness.'

Ecofys estimates that the costs of capacity building for sectoral CDM in Chile – which is likely to be a PMR participant – would be more than US$14 million, rising to US$25 million if sectoral targets were adopted (Vieweg et al. 2009).[4] In other words, Chile would contribute two-thirds of the overall costs of developing a scheme from which it is supposed to be a beneficiary. In the case of China, Ecofys estimates that capacity building for sectoral CDM would cost US$26 million, rising to US$57 million for the implementation of sectoral targets and US$130 million for a scheme linked to national targets.

Each project is expected to last three to five years. The initial US$100 million sought to make the Fund operational is also expected to be spent over a time scale of up to five years. However, '[t]he Partnership itself does not have a sunset clause and will continue to provide support as long as there is demand from countries for market readiness capacity building and piloting' (European Commission 2010a:26).

Avoided responsibility mechanisms

Various rationales have been offered for scaling up carbon markets. The *Synthesis Report*, for example, introduces sectoral carbon markets as 'mechanisms to broaden the scope of mitigation' (UNFCCC 2011a:7). In the context of this debate, it is frequently claimed that the Kyoto Protocol's flexible mechanisms, most notably the CDM, are unable to achieve the levels of emissions reductions needed to stop runaway climate change.

This is undoubtedly the case, although not necessarily for the reasons put forward by proponents of expanding carbon markets. Richard Baron of the International Energy Agency (IEA), for example, points out that the CDM covers less than 1.5 gigatons of carbon dioxide ($GtCO_2$) of electricity production – with claimed 'reductions' of 400–600 megatons of carbon dioxide ($MtCO_2$) – in developing (non-Annex I) countries, out of a total electricity sector that generates 60 $GtCO_2$ in the 2000–2012 period that Baron analyses. The sector alone has seen an 8 percent annual increase in CO_2 emissions. On this basis, Baron concludes that the CDM is 'structurally unlikely to deliver needed mitigation' and that new mechanisms are therefore required (Baron 2010).

Fundamental questions of equity are overlooked here. While the rise in emissions in the developing countries is noted as a potentially alarming trend by the IEA, the historical and present emissions of industrialised countries are not addressed. In detaching emissions trajectories from a broader view of global emissions, the implication is clearly made that climate mitigation actions should be targeted on the developing countries. This fails to deal with the underlying *structural* factors contributing to an increase in emissions in these countries – which include export-led development models that have seen a significant proportion of emissions rise as a reflection of outsourced emissions from Annex I countries (Peters et al. 2011).

The distribution of responsibility for climate action is directly tied to the context in which new market mechanisms are being proposed. Scaling up markets in the developing countries is conceived as a means to draw non-Annex I countries into engaging in more widespread mitigation actions. Such proposals assume a 'high-ambition' world in which industrialised countries take bold actions to cut their emissions domestically. As the UN Climate Change Conferences in Copenhagen and Cancún made abundantly clear, this ambition is resolutely lacking.

Carbon financing: Sending the wrong kind of money to the wrong places

New market mechanisms are being proposed with the aim of pushing an increasing proportion of climate financing through the carbon market. Such a conclusion was, for example, reached by the UNFCCC secretariat when looking at the 'investment and financial flows' associated with climate change mitigation. In 2007, it estimated that US$90–100 billion per year would need to be invested in developing countries by 2030, while the value of the carbon market was estimated at US$10–100 billion (UNFCCC 2007b). On this basis, it concluded that the carbon market 'would have to be significantly expanded to address needs for additional investment and financial flows' (UNFCCC 2007b:par. 6).

Yet carbon market revenues are far from the only financial flows available to address climate change – and, in fact, tend to provide an extremely poor source of financing. Carbon market-related investments are often accompanied by significant flows of money and resources out of the host country that are not accounted for within climate finance frameworks, and can have broader destabilising effects on the financial system. The increasing reliance on private equity, for example, encourages risky investments that are subject to a far higher failure rate than public finance – a poor basis for the infrastructure investments that climate finance purports to encourage. This tends not to show up in the economic modelling surrounding carbon market financial flows. Such models also exclude the significant pressures that carbon markets bring to bear in terms of resource extraction and land expropriation.

Widening the gap

A further problem with new market mechanisms is that they reinforce and possibly exacerbate distortions in how climate financing is distributed – first, by according a greater proportion of money to mitigation efforts than adaptation and, secondly, by concentrating financial incentives in the hands of large corporations in middle-income countries. As the Centre for European Policy Studies puts it,

> There is a risk that credits would most likely end up with already dominant companies in emerging economies, because of, for example, their size, technical and/or political savvy, access to resources and management, and sheer economic weight … This risk is even greater if companies continue to be state-owned or close to the government (CEPS and WBCDS 2008:34).

More importantly, with these new mechanisms targeting major polluting industries and manufacturing for support, local communities would again be the losers – faced with subsidies to significant polluters that often have a long record of disregarding local health and environmental pollution issues (Gilbertson and Reyes 2009:53–87).

Such markets are also a poor means of driving long-term finance. Relatively high transaction costs mixed with significant investment risk make carbon-crediting mechanisms a poor source of such money (Sterk 2010:7). While the market does, in theory, offer a means to address this by the 'forward-selling' of credits, the cure is in many ways worse than the disease – since it implies gambling on yet-to-be issued credits. This speculative market, which mostly involves trades made 'over the counter', not only tends to be a source of short-term profit for financial institutions rather than a stable income stream for project building, but also contributes to the formation of destabilising speculative bubbles (Kill et al. 2010:87–106).

New mechanisms in a contracting market

For now, though, the carbon market looks less like a bubble and more like a busted flush. Even proponents of such markets have noted the dangers of advancing new mechanisms in this context. As the IEA pointed out in January 2010,

> Current estimates show that the supply of credits through scaled-up market mechanisms could be significantly larger than demand … Some observers point to the risk of market flooding, resulting in lower carbon prices and slower mitigation efforts in Annex I countries (Aasrud et al. 2010:118).

These risks continue to increase.

In the immediate aftermath of the Copenhagen conference (COP 15), Bloomberg New Energy Finance, a major carbon market consultancy, estimated that demand for international offsets would reach 4 280 $MtCO_2$ over the eight-year period from 2012 to 2019, equivalent to an average of 530 $MtCO_2$ per year (Turner 2010:96). By way of comparison, Bloomberg estimated the supply of international offsets from existing CDM and Joint Implementation (JI) schemes ranges from 2 480 $MtCO_2$ (310 Mt/yr) to 4 460 $MtCO_2$ (560 Mt/yr).

Fast forward eighteen months, and the estimated demand for carbon credits has fallen even further. The World Bank's *State and Trends of the Carbon Market 2011* estimates a demand of between 2 920 $MtCO_2$ and 3 910 $MtCO_2$ of offset credits for the 2013 to 2020 period (World Bank 2011:66, 63). This range includes an estimated demand of 1 750 $MtCO_2$ of offset credits from within the EU if it sticks with its current 20 percent emissions reduction target (compared to 1990 levels), or 2,550 $MtCO_2$ if the EU adjusts its target to 30 percent. The World Bank optimistically speculates that Australia will start purchasing carbon credits in 2015 and that none of the large surplus of 'hot air' Assigned Amount Units (AAUs, a unit of emissions reductions issued under the Kyoto Protocol) will be rolled over for use by governments attempting to meet emissions reduction targets in the post-2012 period. Its figures, moreover, reflect a 'maximum theoretical demand' (World Bank 2011:66).

By comparison, the World Bank estimates that 2 500 $MtCO_2$ offsets will be generated, with 50 to 70 percent of these coming from CDM projects registered before 2012 (World Bank 2011:67). The reduction in the projected supply of credits factors in the impact of new restrictions imposed by the EU in the third phase of its ETS, which begins in 2013. The EU ETS will restrict the use of CDM credits to those issued by projects registered prior to 2013, with the exception of projects undertaken in Least Developed Countries (LDCs). It will also disallow the use of credits from hydrofluorocarbon (HFC) and nitrous oxide (N_2O) industrial gas projects, which account for 67 percent of the total issued to date (World Bank 2011:48). This reflects the stated strategy of the EU for the future of the global carbon market: restricting the CDM to LDCs and developing new market mechanisms in its place to draw middle-income countries into cap and trade schemes related to binding emissions targets. With 97 percent of demand for carbon credits primarily driven by its ETS, the EU can, to a significant extent, force through its position on the future of carbon markets by means of domestic rule changes, irrespective of international climate negotiations (World Bank 2011:9).

Comparing these supply-and-demand projections shows that even with the EU's rule changes factored in, the World Bank's 'optimistic' estimate still leaves just

400–1 400 $MtCO_2$ of demand that is unmet by the existing CDM in the 2013 to 2020 period – at the low end, just 50 $MtCO_2$ per year. By way of comparison, the emissions from the largest single power plant within the EU ETS (Elektrownia Bełchatów in Poland) are currently almost 90 $MtCO_2e$ per year (European Commission 2011).

The World Bank does not follow through on its exercise to estimate the potential supply of credits from new sectoral market mechanisms. The submissions to the UNFCCC consultation and the *Synthesis Report* are similarly silent on the supply-side question. However, an IEA/OECD study gives an overview of projections as to the potential scale (Baron et al 2009b:16). It estimates that sectoral crediting in the power sector could amount to 465 $MtCO_2$ annually, or 3 700 $MtCO_2$ from 2013 to 2020. Other studies cited by the IEA and OECD project a potential supply of 110–560 $MtCO_2$ annually for a multi-country power sector scheme including China, India, South Africa, South Korea, Mexico, Indonesia, and Thailand; 154–767 $MtCO_2$ annually if it were to cover only the power sector in China; an additional 460–720 $MtCO_2$ annually if the cement sector in China, Mexico, and Brazil were to fall under a sectoral crediting scheme; and 1 $GtCO_2$ if the iron and steel sectors in non-Annex I countries were to do likewise (Baron et al. 2009b:16).

Serious questions therefore need to be raised about the potential demand for the credits generated by new market mechanisms. Without additional restrictions on the use of carbon credits, it is likely that the creation of new market mechanisms would create a surplus of credits that could collapse the price of carbon – undermining the purported rationale of the scheme.

Climate financing: A disappearing act

There are many good reasons why a further collapse in the carbon market would not be mourned by observers interested in addressing climate change equitably and justly. It is, however, worth noting that the creation of new market mechanisms risks a contagion that would spread beyond carbon trading to climate finance more generally.

To appreciate this more fully, the new market mechanisms should be seen in the context of a far broader shift from public to private climate finance. Industrialised countries have disproportionately contributed to causing climate change, and therefore face obligations to tackle it. This is often referred to in the UNFCCC debate under the rubric of 'common but differentiated responsibilities', while climate justice activists refer to it as 'climate debt' (Working Group on Climate Debt 2009). Insofar as the differentiated responsibilities for climate debt imply financial transfers, they have typically been conceived of as involving public money.

Many industrialised countries now face significant debt burdens as a result of bailing out their financial sectors, with the IMF estimating an increase in industrialised country debt-to-GDP ratios of 110 percent by 2015 (IMF 2010). In the face of such projections, these countries are failing to take responsibility for the climate problem and are searching instead for 'innovative' sources of financing, with particular emphasis on the private sector.

Counting carbon market revenues as 'climate financing' conveniently bridges the funding gap on paper, but does so by removing the element of obligation. Private money is assumed to take the place of public investment, yet this is likely to result in a bias towards projects that both have a high risk of failure and that are premised on a high rate of return for (developed world) investors – the full costs of which are not factored in when considering these new forms of financing (Bretton Woods Project 2010). The 'readiness' mechanisms preparing countries to receive such financing tend to involve an economic liberalisation that has a destabilising effect on economies.

This represents a financialisation of climate change mitigation; ironically, the same processes in the global economic arena – an overly financialised economic model – created the public debt crisis now facing industrialised countries.

These measures also introduce a significant risk that the projected carbon market revenues that are being put at the heart of climate finance will fail to materialise. The UN Secretary General's High Level Advisory Group on Climate Change Financing (AGF) suggested that US$30 to US$50 billion per year in climate financing could be achieved from additional carbon market financial flows by 2020 (Secretary General's AGF 2010). However, if the forecasts are re-run using the demand estimates and carbon prices from the World Bank's *State and Trends of the Carbon Market*, the figure is very different: financial flows related to carbon offsets could amount to just US$3.99 billion per year.

On top of the issues of environmental integrity and social justice that invariably accompany the carbon market – the need to compare such figures to the flow of resources and finance from the developing world to the developed world, the increased risk that carbon finance entails, the element of 'double counting' involved in treating carbon credits as mitigation and finance, and the lack of ambition shown by the AGF (and Green Climate Fund's) US$100 billion per year figure – this suggests that a central emphasis on market mechanisms in the provision of climate finance could create a large hole in the figures, with the money failing to turn up.

The carbon market zombies stumble on

New sectoral carbon markets are presented as a means to move beyond the CDM and scale up mitigation actions in the developing world. However, increasing the size of carbon markets is not the same as reducing emissions. The evidence of the CDM to date suggests that offsetting *increases* rather than reduces greenhouse gas emissions. New sectoral mechanisms risk scaling up these failings.

The introduction of new markets in the context of a declining global trade in carbon throws this into sharp focus. If new mechanisms start delivering significant quantities of credits in a market with limited demand for them, the price of carbon would likely collapse. Introducing new markets in a context of unambitious climate action by industrialised (Annex I) countries is likely to undermine both climate change mitigation efforts and flows of climate finance.

These new mechanisms may well help industrialised country governments and corporations to delay meaningful domestic action to reduce their greenhouse gas emissions. An alternative to the standard explanation of the push for new market mechanisms as a means to 'scale up' climate change mitigation efforts, or correct the environmental and equity failings of offsetting under the CDM, is therefore that they are primarily intended to address the corporate competitiveness concerns of industrialised countries. In so doing, they are likely to shift environmental and fiscal responsibility for tackling climate change towards middle-income countries in particular, and countries in the developing world more generally. The attempt to advance new carbon markets within the LCA track of climate negotiations can also be seen as part of a broader project of international climate change 'regime change', embedding carbon trading in this new infrastructure even as the legally binding targets they were initially tied to by the Kyoto Protocol are shed.

The markets themselves seem rather less keen than governments on these new initiatives, however. With EU economies slipping into a potentially deeper financial crisis exacerbated by austerity measures, production is expected to flatline – reducing demand for permits and credits from the utilities and industrial producers covered by the ETS. These 'compliance' buyers already hold a significant surplus of permits, estimated at up to 1.2 billion between 2013 and 2020 (Environment Agency 2010:16). At the same time, EU measures to limit industrial offset credits after April 2013 have led to their dumping onto the international market, precipitating a price collapse (Wynn and Chestney 2011). As we have shown, the overproduction of emissions allowances looks likely to remain a problem, further undermining the environmental integrity of the scheme, and begging the question: why are governments and international financial institutions still pushing for new markets?

Part of the answer rests with institutional inertia – 'new market mechanisms' were tabled when the USA was planning a federal cap and trade market, with an almost tenfold increase in demand compared to the EU ETS. The delays and downscaling of expectations for cap and trade schemes in other industrialised countries are, in part, a response to the failure of legislation on climate change in the US. Whereas carbon markets emerged as a plan B for governments and corporations looking to avoid restructuring their power production or industrial base, the plan A of not legislating on climate change at all has also regained ground (driven on by a climate-sceptic Right in the USA, Australia, Canada, and Japan, in particular). This manifested itself in the Japanese government's rejection of a second Kyoto commitment period at Cancún, which will probably be matched by US intransigence in Durban, where its delegation is unlikely to countenance any significant moves on the climate ahead of an election year.

The push for new carbon markets, by contrast, is being driven by the EU. Going into Durban, the rotating EU presidency is held by Poland, probably the most climate-sceptic government of the 27-member bloc. Within the European Commission, moreover, there is an internal dispute on the centrality of carbon trading – with two significant departments (DG Industry and DG Enterprise) pushing back against it. But the negotiating agenda on new market mechanisms is largely controlled by DG Climate Action, whose lead officials and official negotiators made their careers off the back of promoting the EU ETS.[5] With the centrality of the ETS being part of a broader power struggle in Brussels, it is unlikely that DG Climate Action will easily give up on carbon trading.

The ideological commitment to carbon markets also retains a strong grip. Against a growing body of evidence, the proponents of trading continue to present it as the theoretically optimum means to put a price on carbon, and to suggest that such pricing should be central to action on climate change.[6] This is sometimes allied to the view that new programmatic and sectoral carbon market instruments will serve as 'stepping stones' to a global cap and trade system (European Commission 2010b; Lazarowicz 2009).

The less rosy-eyed among them may realise that such a system would entail a patchwork of rules, triggering a race to the bottom in terms of environment safeguards – although if they do, they are not yet saying so. Either way, we can expect Durban to see a renewed push for the extension of existing carbon markets alongside an increased emphasis on the private sector in climate finance, and an expansion from 'carbon' towards broader biodiversity markets (which may continue through to the Rio+20 summit in June 2012).

A proposal to include Carbon Capture and Storage (CCS) in the CDM will be tabled for agreement in Durban, introducing a potential new subsidy stream for the oil and coal industries (Reyes 2011a). The expansion of carbon offsets is also likely to continue apace, with the World Bank considering the launch of a new Fund for 'soil carbon' credits and other 'climate-smart' measures in Durban. Such proposals are likely to prove a boon for agribusiness but would be of little benefit to small-scale farmers: a BioCarbon Fund-backed pilot project in Kenya anticipates consultants' fees for registering and monitoring the project in excess of US$1 million, while each farmer involved would get around US$1 per year (Sharma 2011).

Reducing Emissions from Deforestation and Forest Degradation (REDD+) schemes remain a key plank of the climate negotiations, too. These have been conceived of as the basis for a forest carbon market from the outset (Heal and Conrad 2005). The fact that most REDD money is currently provided by the Norwegian sovereign wealth fund (and the aid budgets of other industrialised country governments) does little to alter this fact: even if a 'pledge and review' system were to precipitate a collapse of the Kyoto institutions, the Voluntary Carbon Standard (VCS) supported by the World Bank BioCarbon Fund would likely become a de facto basis for a continued voluntary market.

The '+' in REDD+ signals its expansion to encompass a broader array of biodiversity markets, moreover. In the face of potentially low carbon prices, the World Bank (and WWF) have even promoted a Wildlife Premium Market Initiative that would issue additional certificates alongside REDD credits for the protection of 'charismatic species' (Zoellick 2010).

Beyond the formal negotiations themselves, the World Bank and Regional Development Banks continue to proliferate their stock of carbon funds. The World Bank Group owns or administers fourteen such funds, at the latest count. Its most recent moves have included a €68 million refinancing of the Umbrella Carbon Fund to 'bail out' investments that have been badly exposed by the collapse in post-2012 carbon credit prices (*Business Green* 2011). The IFC's new post-2012 Carbon Fund serves a similar purpose.

At a bilateral level, Japan consolidated its rejection of the Kyoto Protocol in Cancún with the creation of a 130 billion yen (US$1.7 billion) bilateral fund to promote Japanese technology exports in return for voluntary carbon credits that Japan would purchase (Young 2011). Its initial projects include coal and nuclear power plants in

Indonesia and Vietnam and a carbon capture and storage project in Indonesia – at odds with the CDM, which currently excludes both. Japan has sought to include these proposals in the LCA negotiations on new market mechanisms (UNFCCC 2011a), but the fund will exist irrespective of the negotiations' outcomes.

The proposals surrounding the PMR highlight numerous further bilateral and development-bank-funded initiatives in the pipeline. The PMR itself is backing emissions trading pilot schemes in two provinces (Guangdong and Hubei) and four municipalities (Tianjin, Shanghai, Beijing, and Chongqing) in China, with the EU similarly offering 'technical assistance' to promote these schemes (NDRC 2011). Turkey is considering establishing a national carbon market as part of a broader strategy aimed at the financialisation of its economy (Undersecretariat of Treasury 2011:4), with technical support from the PMR and UN Development Program. Ukraine is relying on a similar support structure – alongside the EU, German government, and European Bank for Reconstruction and Development (EBRD) – to develop its own national carbon-trading scheme (State Environmental Investment Agency 2011). Mexico is seeking to develop sectoral Nationally Appropriate Mitigation Action (NAMA) crediting in the housing, waste, cement, and transport sectors – with the backing of a range of foreign development agencies and consultancies, as well as the PMR (CICC 2011); Thailand has expressed interest in sectoral (or NAMA) crediting; Morocco seeks support for its CDM Programme of Activities proposals in the waste sector (which are also backed by the Carbon Partnership, another World Bank Fund); and so on.[7]

Durban may agree to create new market mechanisms, or it may remain deadlocked. The carbon market looks set to continue its price slump, with traders gradually losing interest and diversifying attention to a broader range of biodiversity markets. But at a bilateral and national level, encouraged by the EU, the World Bank, and other International Financial Institutions (IFIs) and development agencies, the zombies of new carbon market mechanisms continue to stumble on.

References

Aasrud, A., R. Baron, B. Buchner, and J. Ellis. 2010. Scaled-up market mechanisms – What is needed to move forward? In *Greenhouse Gas Market Report 2010. Post Copenhagen and climate policy: Where global emissions trading goes from here*, ed. A. Kramer, 115–118. Geneva: IETA.

Australian Government. 2010. Australia's fast start finance: Progress report, http://www.climatechange.gov.au/en/media/whatsnew/~/media/publications/international/australias-fast-start-finance-progress-report.ashx.p.10.

Baron, R., J. Newman, A. Aasrud, B. Buchner, K. Inoue, and S. Moarif. 2009a. *Sectoral approaches in electricity: Building bridges to a safe climate.* Paris: IEA and OECD.

Baron, R., B. Buchner, and J. Ellis. 2009b. *Sectoral approaches and the carbon market.* Paris: IEA/OECD.

Baron, R. 2010. Proposals for sectoral crediting, sectoral trading and national appropriate mitigations actions in international negotiations. EPRI Greenhouse Gas Emissions Offset Policy Dialogue Workshop 7 – Sectoral and International Crediting Mechanisms, Washington D.C., February 25, http://www.iea.org/speech/2010/baron_washington.pdf.

Braun, M. 2009. The evolution of emissions trading in the European Union – The role of policy networks, knowledge and policy entrepreneurs. *Accounting, Organizations and Society* 34 (3–4): 499–534.

Business Green. 2011. World Bank puts up €68m to avert post-Kyoto carbon market crash, January 14, http://www.businessgreen.com/bg/news/1937059/world-bank-68m-avert-post-kyoto-carbon-market-crash.

Bretton Woods Project. 2010. The private sector turn, http://www.brettonwoodsproject.org/art-567281.

Center for European Policy Studies and World Business Council for Sustainable Development (CEPS and WBCSD). 2008. *Global sectoral industry approaches to climate change.* Brussels: CEPS and WBCSD.

Environment Agency. 2010. *Report on 2009 EU Emissions Trading System emissions data.* Bristol: Environment Agency.

European Commission. 2010a. 2010 Annual action programme (Part II): Implementing the thematic programme for the environment and the sustainable management of natural resources including energy (ENRTP), http://ec.europa.eu/europeaid/documents/aap/2010/af_aap_2010_dci-env_p2.pdf.

———. 2010b. Linking the EU ETS to other emissions trading systems and incentives for international credits, http://ec.europa.eu/clima/policies/ets/linking_en.htm.

———. 2011. Verified Emissions for 2008-2009-2010 and allocations 2008-2009-2010. April 15, http://ec.europa.eu/clima/documentation/ets/docs/registries/verified_emissions_en.xls.

Gilbertson, T. and O. Reyes. 2009. *Carbon trading: How it works and why it fails.* Uppsala: Dag Hammarskjold Foundation.

Harrison, P. 2011. EU energy plan threatens carbon billions. *Reuters.* May 30, http://uk.reuters.com/article/2011/05/30/us-eu-energy-climate-idUKTRE74T23C20110530.

Heal, G. and K. Conrad. 2005. A solution to climate change in the world's rainforests. *Financial Times.* November 29, http://www.ft.com/intl/cms/s/1/032d0496-610c-11da-9b07-0000779e2340.html#axzz1UdrLzuit.

Interministerial Commission on Climate Change (CICC). 2011. Template for organizing framework for scoping of PMR activities. May 23, http://wbcarbonfinance.org/docs/Mexico_Organizing_Framework_May_30_2011.pdf.

International Monetary Fund. 2010. *Navigating the fiscal challenges ahead.* Washington D.C.: IMF.

Kill, J., S. Ozinga, S. Pavett, and R. Wainwright. 2010. *Trading carbon: How it works and why it is controversial.* Moreton in Marsh: FERN.

Lazarowicz, M. 2009. *Global carbon trading: A framework for reducing emissions*. London: The Stationary Office.

Lohmann, L. 2009. Toward a different debate in environmental accounting: The cases of carbon and cost-benefit. *Accounting, Organisations and Society* 34 (3–4): 499–534.

National Development and Reform Commission (NDRC) China. 2011. Template for organizing framework for Scoping of PMR activities. May 19, http://wbcarbonfinance.org/docs/China_Organizing_Framework_May_19_2011.pdf.

Peters, G., J. Minx, C.L. Weber, and O. Edenhofer. 2011. Growth in emission transfers via international trade from 1990 to 2008. *Proceedings of the National Academy of Sciences* 108 (21): 8903–8908.

Reyes, O. 2011a. Carbon markets after Cancún: Carbon capture and storage in the Clean Development Mechanism. January 20, http://www.carbontradewatch.org/articles/carbon-markets-after-Cancun-carbon-capture-and-storage-in-the-clean-development-mech.html.

———. 2011b. *More is less: A case against sectoral carbon markets*. Barcelona: Carbon Trade Watch, http://www.carbontradewatch.org/articles/more-is-less-a-case-against-sectoral-carbon-markets.html.

Secretary-General's High-level Advisory Group on Climate Change Financing (AGF). 2010. Report of the Secretary-General's High-Level Advisory Group on Climate Change Financing. New York: United Nations, http://www.un.org/wcm/webdav/site/climatechange/shared/Documents/AGF_reports/AGF%20Report.pdf.

Sharma, S. 2011. Agriculture in the spotlight at Bonn climate talks. June 16, http://iatp.typepad.com/thinkforward/2011/06/agriculture-in-the-spotlight-at-bonn-climate-talks.html.

State Environmental Investment Agency (Ukraine). 2011. Template for organizing framework for scoping of PMR activities. May 31, http://wbcarbonfinance.org/docs/Ukraine_Preparations_Status_May_31_2011.pdf.

Sterk, W. 2010. New mechanisms for the carbon market? Sectoral crediting, sectoral trading, and nationally appropriate mitigation actions. *JIKO Policy Paper 4/2010.* Wuppertal Institute for Climate, Environment and Energy. June, http://www.jikobmu.de/english/background_information/publications/doc/961.php.

Third World Network. 2009. Understanding the European Commission's climate communication. Penang: Third World Network.

Turner, G. 2010. Copenhagen and the implications for supply and demand dynamics in the global carbon market. In *Greenhouse gas market report 2010. Post Copenhagen and climate policy: Where global emissions trading goes from here*, ed. A. Kramer, 94–97. Geneva: IETA.

Undersecretariat of Treasury, Turkey. 2011. Partnership for Market Readiness (PMR) expression of interest in participating in the PMR, http://wbcarbonfinance.org/docs/Turkey_EoI.pdf.

UNFCCC. 2007a. Bali Action Plan, http://unfccc.int/files/meetings/cop_13/application/pdf/cp_bali_action.pdf.

———. 2007b. *Investment and financial flows to address climate change*. Bonn: UNFCCC.

———. 2011a. Synthesis report on information on various approaches in enhancing the cost-effectiveness of, and promoting, mitigation actions: note by the secretariat. March 30. FCCC/AWGLCA/2011/4.

———. 2011b. Market-based and non-market-based mechanisms: Views on the elaboration of market-based mechanisms. Submissions from Parties. March 21. FCCC/AWGLCA/2011/MISC.2.

Vieweg, M., C. Doets, M. Torres, S. Wartmann, P. Oliver, D. Yener, E. Cruz, E. Dalenoord, and N. Valero Esteban, 2009. *Linking developing countries to carbon markets: Cost assessment of capacity building requirements*. London: Ecofys U.K.

Working Group on Climate Debt. 2009. We demand the enforcement of the payment of climate debt. World Conference of Peoples on Climate Change and the Rights of Mother Earth. April. Cochabamba.

World Bank Climate Finance Unit. 2011a. Partnership for Market Readiness (PMR): Update and preliminary insights. March, http://www.oecd.org/dataoecd/55/19/47517775.pdf.

———. 2011b. Partnership approves grants for eight carbon market initiatives. June, http://go.worldbank.org/UOGJZTB2Z0.

———. 2011c. Chair's summary of first meeting of the partnership assembly. June 15, http://wbcarbonfinance.org/docs/PA1_Chair_s_Summary_June_15.pdf.

World Bank. 2011. *State and trends of the carbon market 2011*. Washington, D.C.: World Bank Group.

Wynn, G. and N. Chestney, 2011. Carbon offsets near record low, worst performing commodity. *Reuters*. August 8.

Young, T. 2011. Japan promises $1.6bn clean tech exports investment. *Business Green*. January 6, http://www.businessgreen.com/bg/news/1935223/report-japan-promises-usd16bn-clean-tech-exports-investment.

Zoellick, R. 2010. World Bank group president Robert B. Zoellick remarks at Cancún summit on REDD+. World Bank Group. December 7, http://web.worldbank.org/WBSITE/EXTERNAL/COUNTRIES/AFRICAEXT/0,,contentMDK:22786625~menuPK:258662~pagePK:2865106~piPK:2865128~theSitePK:258644,00.html.

Endnotes

1 I would like to gratefully acknowledge the support of the Isvara Foundation, which made possible the writing of this article. Thanks to Joanna Cabello, Tamra Gilbertson, and Larry Lohmann for comments and discussion.

2 For more details on these proposals, including a critique of their environmental failings, see *More is Less: A Case Against Sectoral Carbon Markets* (Reyes 2011b).

3 The Kyoto Protocol assigned responsibility for reducing international aviation and shipping emissions to the ICAO and IMO. A number of countries and blocs have expressed frustration at the inability of these institutions to reach agreements on climate change. The EU has legislated for the inclusion of aviation in its ETS starting from January 2012 and is also considering including ships that dock in the EU in the scheme.

4 The figures quoted for Chile are a like-for-like comparison with the capacity building estimates given by the PMR – both assume sectoral crediting and exclude implementation costs.

5 DG Climate Action Director-General Jos Delbeke; Head of Policy Coordination Peter Zapfel; and Commissioner Hedegaard's Chef de Cabinet Peter Vis were key figures in the "policy network" that promoted the creation of the EU ETS. For more details, see Braun 2009.

6 For extended critiques of this position, see Gilbertson and Reyes (2009) and Lohmann (2009).

7 The World Bank is publishing PMR background documents at: http://wbcarbonfinance.org/Router.cfm?Page=PMR&FID=61218&ItemID=61218&ft=DocLib&dl=1&ht=63206.

Chapter 10

Kleptocratic Capitalism, Climate Finance, and the Green Economy in Africa[1]

Yash Tandon

One of the most difficult issues of our times is the question of sustaining growth and development in Africa whilst at the same time protecting the environment and ensuring that growth benefits entire communities and not just a small minority. There are only two points I wish to make here. One is that Africa's development and growth model is seriously flawed. It has not translated into enhancing people's welfare over the last 40 to 50 years. And the second is to caution African countries and the African Union against outsourcing policy issues, especially those relating to international negotiations – such as the case with climate change – to experts from outside Africa.

In a recent paper, the UN Economic Commission for Africa argues that despite high growth rates in Africa, there has been no improvement in employment and welfare of ordinary people. The paper gives Africa's commodity export dependence as the primary reason. I agree, but the main reason, in my view, is the global system of production and exchange of which Africa is a part (AUC/UNECA 2011).

The global system of production of wealth and its distribution is characterised by kleptocracy, primitive accumulation, and dispossession. Economists call it 'rent-seeking', and they justify this with the argument that surplus from the rural and agricultural areas is needed in order for Africa to grow and industrialise – that this is how they did it in the West. That is historically true.

However, the early model of competitive capitalism of the 19th and the 20th centuries is not applicable to Africa today because we live in a different world. It is a world of *kleptocratic capitalism.* Some leftist intellectuals call it *financialised capitalism* – a system where finance is king and production takes second place. But this term is limited only to its predominant economic characteristic. *In essence, in political-economic terms, kleptocratic capitalism is a system of economic production and exchange; the creation of fictitious wealth without going through the production of real wealth; and political governance controlled by looters and daytime robbers.* It is rent-seeking by the rich nations, and within each nation by the rich economic and power elite. At the opposite polar end, this creates the dispossession and disempowerment of the masses.

This, when everything is said and done, is at the root of the rebellions by the masses in Tunisia, Egypt, Libya, Bahrain, Yemen, and other Arab and North African countries. This is a current that could swell into a tsunami tidal wave to other parts of Africa – if not today, then tomorrow or the day after. It is inherent within the very dynamics of kleptocratic capitalism's contradictions.

At the global level we have Ponzi schemes that create wealth out of thin air – money made without going through real production, money from fraudulent deals and speculation. The Madoff investment scandal in the US is a good example. Bernie Madoff is now in jail, but for a long time he was the chairman of NASDAQ,[2] a stock market that deals mainly with industrial and technology shares. But he is not the only one. *The global banking system is itself, by its nature, a huge Ponzi scheme.*

What is happening in Europe, for example, is a good indicator: German and French bankers buy Greek bonds guaranteed by the Greek government – ostensibly to bail out a bankrupt Greek treasury. But in the process, the banks exert pressure through their governments, the European Commission, the European Central Bank, and the International Monetary Fund (IMF), to impose severe austerity measures on the people of Greece in order to ensure that the Greek government does not default on the loans. If it does, it will put into jeopardy the EU currency system, and possibly the entire European project that is largely pushed by European corporate and finance capital. The people of Greece have to suffer so that the 'system' survives. The system's survival is more important than the well-being of the people. The term 'systemic risk' has become part of the vocabulary of economists since the 1997–98 financial and economic meltdown. Ireland was an aid donor to Africa only yesterday; today it is a beggar nation – beggared by the Ponzi-like global banking system.

Globalised capital (the so-called foreign direct investments, or FDIs) and its several manifestations – banks, insurance companies, shipping agents, commodity speculators, wholesale traders, chain retailers, etc. – are in league with the local economic and power elites in the recipient countries, and their god is 'accumulate, accumulate, and accumulate'. Even China and India are not free from this virus – displacement of people from land is creating staggering problems in these countries. These new capitalist countries are still significantly underdeveloped from the perspective of science and technology. The battles over intellectual property in the World Intellectual Property Organization (WIPO) and in the industrial-corporate world testify to this reality. Power in the global banking-financial market lies in the boardrooms of a dozen or so big players such as JPMorgan Chase, Citigroup Inc., Bank of America Corp., Morgan Stanley, Goldman Sachs, and Merrill Lynch. They operate in an unregulated market – i.e., largely outside of national control, not even that of the USA. Like an octopus, they have their tentacles everywhere, including in China, India, Brazil, and South Africa.

Africa undermined

Africa has been one of the major resource providers of global kleptocratic capitalism. By means of so-called development aid and FDIs, Africa's resources are exploited to sustain the system. In real terms, the investments and aid are a gigantic credit system that creates a mountain of debt that Africa has been paying in the form of transfer of real values – coffee, cocoa, cotton, cobalt, platinum, gold, chromium, manganese, uranium, titanium, and other valuable minerals.[3] In a recent article by Kandeh Yumkella, Director-General of the UN Industrial Development Organization (UNIDO), and Rob Davies (2011), South Africa's Minister of Trade and Industry, the authors say that Africa has 80–200 billion barrels of hydrocarbon reserves, but most of these are exploited by global corporations. Or in the case of African titanium sand, one metric ton brings Africa about US$100 in export revenues, whereas a metric ton of titanium alloy brings US$100 000 to countries outside Africa (Yumkella and Davies 2011) – a ratio of 1:1 000.

The gross exploitation of Africa's resources is underpinned by a global credit system run by the World Bank, the IMF, and the aid industry. Development aid is a charade (Tandon 2008). What the IMF and the German-French-dominated banking system have been doing to the peripheral European countries (Greece, Ireland, Portugal) in recent years, is exactly what the IMF, the World Bank, and the so-called donor community have been doing to Africa for the last 50 years.

Africa has its own domestic over-consuming power and economic elite (the plutocrats) in league with their imperial overlords – the bank-robbers and global corporations – that exploit the masses. One of the manifestations of their greed is the massive land grab that we are witnessing today, much of it done by these domestic plutocrats. In addition, many African governments are selling off or leasing agricultural lands to foreign investors from Europe, the US, India, China, the Gulf States, and further afield. There is a rush for all of Africa's resources, not just land, but also Africa's forests, oil, gold, and diamonds.

The price of this intense exploitation is paid by the ordinary people. In recent court cases in South Africa, for example, tens of thousands of former mineworkers received little or no compensation for occupational lung diseases due to working in asbestos mines and other kinds of toxic environments. Hundreds of thousands of rural African people are displaced and dispossessed to make space for domestic and foreign land-grabbers – often, ironically, to grow food 'for the poor' – using agro-chemicals or the magical biofuel, jatropha. These climatically displaced refugees (CDRs) are swarming the rural countryside and peri-urban areas across Africa.

AGRA (Alliance for a Green Revolution in Africa) is one example of this kind of exploitation that is encouraged by mainstream African economists and power elites. Under the guise of providing Africa with climate-sensitive food crops and flowers, Rockefeller and Gates foundations-funded AGRA (with the blessings of former Secretary General of the United Nations, Kofi Annan) is pushing agro-chemical crops using multi-genome patents. Their objective – or at least the end result – is plain to see: the control over Africa's plant biomass to generate super-profits for the mega-chemical and seed corporations.

From Mali to Mozambique, small peasant farmers are resisting the takeover of their lands and meagre means of sustenance. But they are scattered and too weak in political organisation to mount an effective resistance. When the 'Arab spring' hits the cities of these countries – as these uprisings inevitably will – the displaced and disempowered millions may well enlist in droves into rebel armies to remove the neo-colonial dictators of Africa from their perched thrones.

This, in brief, is the first point. Africa is run by a global kleptocratic system that enriches a minute number of economic and power elite in Africa along with the global bankocrats and corporatocrats at the one end of the pole while it impoverishes the masses of African people at the other end of the pole. Economists call this 'rent-seeking', but it is simply looting.

What can be done? Though it is too vast a subject to answer comprehensively, I will give two contrasting models that exemplify polar conceptions on how to go about addressing the challenges Africa faces. One is South Africa's second Industrial Policy Action Plan (IPAP-2) based on 'market-based policy measures', mega-projects such as the Coega complex near Port Elizabeth, and carbon trading. This is the route of the trodden past – its outcome is predictable. The second is the pledge the government of Rwanda made at the ninth session of the UN Forum on Forests to alleviate poverty in forest communities with a 25-year plan to tackle ecosystem degradation and improve rural livelihoods.

What is significant about the Rwandese concept is its dual objective of saving both the forests and the forest communities. For the environmentalists, forests are simply biomass that on the one hand provide fuel, and on the other hand function as carbon dioxide-absorbing 'lungs' to help counter global warming. But besides the forests, there are also forest dwellers. *The challenge is to save the forests and the forest communities, the people as well as the environment.* Those in African governments and parliaments who are sensitive to the welfare needs of the people, must support people's movements that take on this dual challenge. One example of this is the ROPPA (Network of Farmers' and Agricultural Producers' Organisations of West Africa), which coordinates and strengthens a number of rural women's associations working towards saving the communities as well as the environment.

From COP 17 to Rio+20 and beyond

My second point is of more immediate concern. As I write in August 2011, there are only three months left between now and the 17th Conference of the Parties of the United Nations Framework Convention on Climate Change (UNFCCC), or COP 17, in November 2011, and less than a year to Rio+20 (the United Nations Conference on Sustainable Development, or UNCSD) that will take place in May 2012 in Brazil. In this short period, African governments and activist civil society organisations might draw lessons from the 2009 COP 15 in Copenhagen and the 2010 COP 16 in Cancún. Both these conferences were followed by widespread dissatisfaction on the part of those striving to save the environment as well as the communities that live off what nature provides them.

For the purposes of this essay, I wish to focus on just one lesson. And this is that Africa needs to be wary of the use of finance (or so-called development aid) by the industrialised countries to divide and rule the developing countries. Globally, if there is a near-clear divide between developed and developing countries, it is on the question of climate change. Until the 1990s, most scientific research and diplomatic negotiations on global warming focused on emission mitigation. The language of adaptation first emerged at the Kyoto Protocol in 1997, primarily as a result of pressure from the developing countries. The UNFCCC, which hitherto had focused on mitigation, now recognises the significance of adaptation and the historical responsibility of the industrialised countries to compensate the developing countries for the damage they have caused to the environment during their period of industrialisation. This historical responsibility is reflected in the UNFCCC's treaty provisions that oblige industrialised countries to provide new and additional financial flows (as well as technology transfers) to the developing countries to support the latter's costs for implementing UNFCCC measures and to undertake climate adaptation. The Kyoto Protocol endorsed this principle by placing these two categories of nations with common but differentiated responsibilities in, respectively, Annex 1 and Annex 2 categories.

At the international level, African countries are members of the Group of 77 and China (G77+China), which forms the primary negotiating group for the developing countries in climate change negotiations. This group also includes members of the Organization of Petroleum Exporting Countries (OPEC). With such a differentiated group, it is natural that there should be differences amongst them on their concerns and priorities. The industrialised countries take advantage of these differences in order to divide and dictate to the developing countries the terms of the climate change negotiations. What makes Africa vulnerable is its dependence on the West for development aid and technical experts.

One significant illustration of this is the manner in which the industrialised nations used money and 'technical assistance' as a means of ensuring an outcome at COP 16 in Cancún after they had failed to do so at COP 15. Europe and the US mounted a coordinated offensive to break the ranks of the countries of the South. Some of this was quite overt and open, for example, through the use of 'development aid' and other financial incentives. Other initiatives were covert and secretive, such as the use of US spy networks – exposed, partially, by *WikiLeaks*.

The biggest bribe is the US$100 billion per year in finance for adaptation and low-carbon development to poorer nations by 2020. This is just pie in the sky. Africa should not hanker after it, for even if it materialises, it would be so firmly ring-fenced with conditionalities that it would auction away the sovereignty of African nations at the altar of 'Green Capitalism' or 'Good Governance'. In early 2011, for example, the US withheld about US$350 million in grant money for Malawi disbursed through the Millennium Challenge Corporation to improve the efficiency of the country's energy sector on the grounds that Malawi had failed to observe governance and human rights issues (Chikoko 2011). As for the annual fund of US$100 billion, the fact that it will be administered by the World Bank strongly suggests those funds will be similarly ring-fenced.

In a February 2011 report, 'Storm On the Horizon? Why World Bank Climate Investment Funds Could Do More Harm Than Good', Eurodad (a network of 57 NGOs from nineteen European countries) (Honkaniemi 2011) argues that the World Bank is not the best-placed institution for a legitimate and development-friendly climate-finance architecture for the future. In general, many European NGOs are sympathetic to the concerns of Africa. My own experience working with a number of them on the issue of Economic Partnership Agreements (EPAs), which Europe is trying to foist on Africa, has been very positive. This said, it is important that African countries are self-reliant on matters related to policy issues, especially when it comes to negotiations in the global system. It is an easy (and cheaper) way out to outsource policy advice to 'experts' from Western NGOs on the grounds that African countries do not have the experts or the money to finance them.

This is relevant in relation to climate change negotiations. As we move to COP 17 and Rio+20, it is important that the African Union Commission builds its own network of experts to advise African countries on the technical and political issues that are likely to emerge in the months ahead. It is in this light that I need to caution Africa against the processes being put in place by several interested parties in the West to offer 'technical advice' to 'poor' African countries.

One such measure is the Climate and Development Knowledge Network (CDKN) initiative. The CDKN, founded in March 2010, is a consortium of consultancies and think tanks that helps decision-makers in developing countries design and deliver 'climate-compatible development projects'. CDKN claims that it can help African countries to maximise their opportunities to tap climate finance and build their capacity to manage these funds. It will also support them in assessing climate risks and vulnerabilities, as well as provide legal advice through the Legal Response Initiative (LRI).[4] The CDKN offers 'real-time, free legal advice to climate negotiators'. This makes it suspect. Why should it offer 'free' advice? What is its agenda? Who finances it? Is it not unfair to wonder whether somebody along the chain benefits from this 'free' advice? Nothing comes free.

The CDKN consortium includes the global accounting and business consulting giant PricewaterhouseCoopers; the Overseas Development Institute (ODI), an international development- and humanitarian issues think tank headquartered in the UK; Ecuador-based Fundación Futuro Latinoamericano (FFLA), an organisation promoting sustainable development, formed in response to the 1992 Earth Summit in

Rio; SouthSouthNorth, a network of organisations focused on poverty reduction in the context of climate change with offices in South Africa, Brazil, Tanzania, Mozambique, and Indonesia; the Nepal-based private organisation and development consultant LEAD International; and the UK-based International NGO Training and Research Centre (INTRAC). I know some of them well from previous interactions. The ODI, for example, advertises itself as an 'independent think tank on international development and humanitarian issues'. From my own personal experience with the ODI on matters related to development aid, trade, and Economic Partnership Agreement negotiations, I can say without a moment's hesitation that it is really an arm of British foreign policy – the 'soft arm' of British imperial diplomacy whose 'strong arm' comprises instruments of force, including sanctions and war.

Policy making is not something that African governments should outsource to anybody from outside Africa. It is best for Africa to develop its own expertise rather than depend on outside help – not only on negotiations on climate change, but on all matters of vital policy concern to Africa. It is, of course, understandable that individual African countries may have limited resources to create think tanks and research institutions. But they can take advantage of bigger institutions of which they are members. The most important collective organisation is, of course, the African Union. But there are others. For example, African countries are significantly represented in the South Centre, which was created in 1995 under the inspiration of Third World leaders like Julius Nyerere and Mahathir Mohamad. The Centre is well placed to provide technical expertise on a range of issues from trade negotiations to intellectual property rights, finance for development, and, yes, climate change. The South Centre, for example, made a significant contribution to the African Union Commission's position on the EPA negotiations in a meeting the African Union Commission organised in Kigali in November 2010.

Conclusion and postscript

Real knowledge comes not from information but from a thorough and deep understanding of Africa's situation. There is a vital and strategic distinction between information and understanding (*verstehen*) – an interpretive or participatory examination of social phenomena. *The fundamental reality of Africa is that it is integrated into a global system of kleptocratic capitalism characterised by primitive accumulation or rent-seeking by the rich nations and within each nation by the rich power elite. This creates the dispossession and disempowerment of the masses*. The present phase of the evolution of capitalism is caught up in its own contradictions, but capitalism is not about to disappear. It is a long road. In facing the challenges of the demands of a Green Economy that ensures a sustainable Africa, Africa has to balance the human rights and needs of the masses of its people with the imperative of protecting Africa's environment. Africa should not hand over policy matters and negotiating strategies to outside 'experts', however benign they might appear. Above all, Africa must build a common united position on climate negotiations, in alliance with the other developing countries, leading up to the COP 17 and then Rio+20. The African Union and the South Centre can play a significant role in leading towards Rio and beyond.

There are larger issues at stake than just the negotiations at COP 17 and Rio+20. At COP 16, where no binding international agreement was made on reducing emissions, Bolivia had taken the courageous stand to challenge the whole concept behind the negotiations, and to raise the vital question of the rights of Mother Nature. The Bolivian ambassador to the UN, Pablo Solón Romero (2011), has since then made presentations at various international forums to elaborate on this revolutionary concept. Space does not allow elaboration. At the time of COP 16, Bolivia was isolated. But since then, Bolivia's position that the Cancún negotiations of December 2010 were not a step forward but rather a step back, has been endorsed by the Group of 77, comprising of 131 developing nations including China.

In the meantime, civil society organisations the world over engaged in the climate change negotiations, have expressed anxiety that some of these technical and broader issues might not be seriously addressed at COP 17. The Climate Justice lobby in South Africa has voiced its concern at the lack of moral and political leadership in the country that is going to host COP 17, and that Trevor Manuel, South Africa's former finance minister and a known neoliberal enthusiast, was chosen as co-chair of the Green Climate Fund (Bond 2011).

The COP 17 conference is likely to be overwhelmed by other, more immediate and urgent developments. Among these are the deepening economic and financial crisis engulfing the developed world; the possible meltdown of the European Union; the global effects of the downgrading of the US credit rating; the war in Libya; the killing of more than 70 people in Norway by a psychopath who turned out not to be an Islamic terrorist (as the media had initially suspected) but a blue-eyed Norwegian; and the tumultuous rising of the marginalised populations within the developed world, including most recently (August 2011) in some of the bigger cities of the United Kingdom. These might (indeed are likely to) have a cumulative effect of downgrading the issues related to climate change in the hierarchy of priorities for the international community as we approach COP 17.

And yet, in times like this it is important that ordinary citizens of the world take a step back and look at the pandemonium and chaos that is engulfing the global kleptocratic capitalist system, and consider the climate change negotiations not in isolation but as part of the broader and deeper malaise that we face as humanity. It is a humanitarian crisis. It is a deeper crisis of which the utter failure of the global political elite to provide moral and political leadership on climate change is only one – albeit its most shaky and unstable – aspect. If cynical politicians and global bankocrats exploit the system for their own wealth and power accumulation, this does not invalidate people's legitimate calls for justice, democracy, human rights, environmental security, and peaceful resolution of conflicts. It is a complex struggle, of course. But complexity is no reason for complaisance.

References

African Union Commission and United Nations Economic and Social Council Economic Commission for Africa. 2011. *Governing development in Africa: The role of the state in economic transformation*, issues paper. Meeting of the Committee of Experts of the 4th Joint Annual Meetings of the AU Conference of Ministers of Economy and Finance and ECA Conference of African Ministers of Finance, Planning and Economic Development. Addis Ababa, Ethiopia. March 24–27.

Bond, P. 2011. The banksters and the Climate Fund. *Counterpunch*. April 25, http://www.counterpunch.org/bond04252011.html.

Solón, P. 2011. We cannot command nature except by obeying her. Speech by Ambassador Pablo Solón, Permanent Representative of the Plurinational State of Bolivia to the United Nations, on the Occasion of the General Assembly Interactive Dialogue on Harmony with Nature, New York. April 20, http://boliviaun.net/cms/?p=2014.

Chikoko, R. 2011. US follows Germany to withhold aid to Malawi. *Africa Review*. February 7. http://www.africareview.com/News/US%20holds%20back%20Malawi%20aid%20over%20gay%20and%20media%20laws/-/979180/1103218/-/97rtg2z/-/index.html.

Honkaniemi, N. 2011. Storm on the horizon? Why World Bank Climate Investment Funds could do more harm than good. *Eurodad*. February 10. http://www.eurodad.org/whatsnew/reports.aspx?id=4395.

Tandon, Y. 2008. *Ending aid dependence*. Oxford: Fahamu Books.

———. 2011. Reflections on the Norwegian tragedy. *Pambazuka News* 544. August 11, http://www.pambazuka.org/en/category/features/75606.

Yumkella, K. and R. Davies. 2011. South of the revolution. Project Syndicate, Trademark Southern Africa. May 13, http://www.trademarksa.org/news/yumkelladavies-south-revolution.

Endnotes

[1] This paper is a slightly edited version of a lecture delivered at the Go Green Afrique: Africa Fair And Conference, organised by Initiative Africa in Addis Ababa, June 23, 2011.

[2] NASDAQ is an acronym for National Association of Securities Dealers Automated Quotations. According to Wikipedia, it is 'the largest electronic screen-based equity securities trading market in the United States and second largest by market capitalization in the world'. The National Association of Securities Dealers set it up in 1971, and as of January 13, 2011, it had 2 872 listings.

[3] Africa holds 90 percent of the world's deposits of cobalt, 90 percent of its platinum, 50 percent of its gold, 98 percent of its chromium, 64 percent of its manganese, 33 percent of its uranium, and 80 percent of its columbite-tantalite (Yumkella and Davies 2011).

[4] The LRI provides legal support and training for Least Developed Countries (LDCs) in the UN Climate Change talks. The LRI was set up in mid-2009 by attorneys from Oxfam GB, World Wildlife Fund U.K. as well as several practicing lawyers in London. See http://www.legalresponseinitiative.org/index.html.

INDEX

A

B

C

D

E

F

G

H

I

N

O

T

U

V

W

X

Y

Z